Contents

List of Illustrations

Preface

Introduction

Chapter 1	**Central Cities**	1
Chapter 2	**Suburbs:** **Teleworking Helps the Affluent**	75
Chapter 3	**Exurbs: New Opportunity Beyond The Suburbs**	105
Chapter 4	**Nightmares and Dreams**	159
Chapter 5	**In Conclusion**	205

TELEWORKING AND URBAN DEVELOPMENT PATTERNS

Goodbye Uglyville
Hello Paradise

Melvin R. Levin

University Press of America,® Inc.
Lanham • New York • Oxford

Copyright © 1998
University Press of America,® Inc.
4720 Boston Way
Lanham, Maryland 20706

12 Hid's Copse Rd.
Cummor Hill, Oxford OX2 9JJ

All rights reserved
Printed in the United States of America
British Library Cataloging in Publication Information Available

Library of Congress Cataloging-in-Publication Data

Levin, Melvin R.
Teleworking and urban development patterns : goodbye uglyville, hello paradise / Melvin R. Levin.
p. cm.
Includes bibliographical references.
1. Telecommuting—United States. 2. Suburbs—Economic aspects—United States. 3. Central business districts—United States—Case studies. I. Title.
HD2336.35.U6L48 1998 307—dc21 98-16374 CIP

ISBN 0-7618-1117-6 (cloth: alk. ppr.)
ISBN 0-7618-1118-4 (pbk: alk. ppr.)

∞™ The paper used in this publication meet the minimum requirements of American National Standard for information Sciences—Permanence of Paper for Printed Library Materials, ANSI Z39.48—1984

For Larcom McIntosh
March 1, 1927 -November 25, 1997

Illustrations

To Live and Die in D.C.	48
Gated Community	51
Percent Living in Suburbs and Metropolitan Area	79
Cathy Cartoon	96
Cathy Cartoon	131
Dilbert	177
Doonesbury	178

Preface

An analysis of the probable impact of telecommunications on urban development is necessarily an exercise in crystal ball gazing. This prognosis may be a bit less fanciful because it draws on a good deal of prior research. In particular, I have dipped into three previous works, *Ending Unemployment, Planning in Government* and *Outside Looking In* between 1984 and 1993.

Since this volume was produced sans graduate assistance or other professorial emoluments, the analysis and conclusions were attributable solely to me. I do wish to thank the University of Maryland and the Planning Department for the indispensable and invaluable gift of time to pursue this research. I am also grateful for comments and advice from my colleagues, Sidney Brower and Marie Howland.

There was also critically important help in one crucial area: my editor and typist, Jo McIntosh, with extradordinary patience and good humor, pruned out esoterica whose resonance was limited to persons with excessively long memories and a predilection for out-of-the-way reading. What remains, I trust, is plain text.

Melvin R. Levin
College Park
University of Maryland

Introduction

The diffusion of telecommunications as a partial substitute for the traditional journey-to-work is likely to reinforce trends toward social and geographic fragmentation that have been operative in the past two generations. Specifically, we can look for a transformation of many central cities into a niche market as people with enhanced choices--about 20 per cent of the labor force--elect to live away from undesirable places and people in low-crime communities with high quality schools. We can anticipate an amenity competition among upscale suburbs and exurbs along with a boom in selected recreation areas and in the college towns that offer every attraction for the brain worker: amenities, schools good government and ambience.

Hope for central cities? One beacon is immigrants, frugal, sober, disciplined at work and school and possessed with entrepreneurial skills and a bent toward home ownership. Urban neighborhoods serving as magnets for immigrants have reaped the rewards of a partial population replacement, functional people for the dysfunctional. The open question: Will the immigrants who make it into the mainstream remain in the cities or will they join predecessor populations by moving to the suburbs?

A dark side to the new technology? Dungeons and Dragons x 100. Virtual reality will make it easy and tempting for the troubled adolescents and marginal adults to become super couch potatoes lost to the give-and-take and occasional hard knocks of real life. We can foresee a golden age of therapy, twelve-step support groups and rhetorical hand-wringing as major components of life in the coming millennium. For some people the defining moment is a car breaking down in a blizzard on the way to work, the third burglary or perhaps a mugging. For others, it is the long, long crawl to the office when the soothing tape deck can't compensate for an over-stressed bladder. Cold, crime, congestion. All have a jarring impact on the exasperated, the frustrated, the frightened and the just plain angry.

They wonder: can we stretch our vacation at a balmy resort by weeks or months when the wind-chill factor at home hits 40s below and lingers? Can we avoid the morning rush hour? Could we live in a green place far out in the country, safe from street thugs and the need for three deadbolts and a stranger-averse dog?

A few years ago, the answer would have been "not likely." Work was cemented in an office, nine to five; the daily commute only a half hour off-peak but close to an hour drive in rush hour traffic. Now new technology and new approaches may bring liberation. The new technology is the computer.

Today employers throughout the nation are encouraging--and sometimes forcing--people to work at home (or at least out of the office) because through the computer they can save on employee office and parking space and on time formerly spent commuting that can be put to better use. At the same time employers can retain valued homebound employees while closely supervising computer-based output.

How Many, How Much on the Way?

There are two ways to look at the impact of telecommunications technology. The narrow view is to focus on telecommuters. The broader perspective includes the self-employed teleworkers who use this technology as an adjunct to the mail and telephone.[1]

The U.S. Department of Transportation estimates that there were two plus million telecommuters in 1992 who were telecommuting on an average of one to two days per week. This represented only 1.6 per cent of the labor force, far below other estimates.

By 2002 DOT predicts that the total number of telecommuters will reach 7.5 to 15.0 million people telecommuting an average of three to four days per week. This amounts to 5 to 10 per cent of the labor force in 2002.[2] Another estimate is that the figure will climb to 11 million by 2000.

If self-employed workers are included, these numbers rise by 25 to 30 per cent or about one worker in every nine or ten.[3] A more optimistic estimate puts the upper level at 25 million by 2002 or possibly one in six workers.[4]

Another estimate, based on a combination of government and corporate statistics is that 39 million people (roughly a third of the work force) already do some work out of their homes. Of this total almost 24 million operate home businesses, half full-time, half part-time. An estimated 6.6 million are home telecommuter employees who work at the office from half-a-day to two days a week. The remainder 8.6 million, work at home after office hours. Many are only a short distance away from some telecommuting.[5]

How fast is the number of teleworkers growing? The Southern California Telecommuting Partnership puts the nine-month gain in Southern California in 1996 at an astounding 11 per cent. There is a special factor in this area: "Nine out of ten people who started telecommuting after the earthquake still do it."[6]

How Much? How Many? Other Estimates

As the data suggest, there is no consensus on the number of people who work full-time or part-time at home let alone those who avoid the downtown commute by working in small suburban office centers. In mid-1996 a *New York Times* story reported that:

> As many as 40 million people work at least part time at home with about 8,000 home-based businesses starting daily.... ...some 1.5 million claimed home office deduction...on their tax returns for 1993.[7]

On closer inspection, we find that the media image of the home office worker as a refugee from the commute to corporate headquarters is inaccurate. The data suggest that they represent only a third of the total. There are:

- 9 to 14 million telecommuters defined as persons who work from homes on a regular basis (at least two days a week) for an outside company

- from 10 to 12 million home based workers or those who run businesses from their homes,
- 12 to 16 million independent contractors who work for multiple companies.[8]

The data suggest that the emerging industrial and office parks that create tax base and jobs are generated out of homes, garages and small office buildings, a good reason for the amenity competition to attract professional/entrepreneur populations. The ten places with the highest percentages of residents who work at home include cities as well as upscale, high tech suburban areas ranging from San Diego and Manhattan to Beverly Hills, Bethesda, Berkeley, Austin, Greenwich, Ct., Santa Monica and Calabasas/101 Freeway, CA. The percentages range from 5.2% to 9.4% of the work force. But this does not mean that these people are all out of the commuting loop. Only about two-fifths are either full-time self-employed or do all of their work from home. The majority are part-time self-employed or employees who do work at home after hours.[9] Hence the small number of claimants who can meet stringent IRS criteria for office deductions.

The ranks of telecommuters--or more accurately teleworkers (if we include the self-employed)--have been burgeoning thanks to a basic change in the U.S. employment pattern. This is the shift to temporary and part-time work that reached record levels in 1994. More than one-fifth of the nation's work force--24.4 million Americans--had only part-time or temporary work. Construction, mining and manufacturing where wages averaged from $530 to $630 a week, were shrinking, while service and retail businesses where the pay averaged from $200 to $370 a week were growing.[10] This may be bad news for many once secure workers.

Time magazine adds its version of a total number: 3 million U.S. employees (as of 1995) telecommuting full or part-time; the number growing at a phenomenal 20 per cent per year.[11] Clearly, a 10 or 20 percent annual gain in telecommuters is unlikely --unless we take into account the part-timers, who make up a major share of the total. This estimate is based on a minimalist definition:

> ...someone who works from home as little as one day a month, during usual business hours...that telecommuters work at home an average of 39.6 hours a month -- or only about 12 weeks a year-- means that most work is still done at the company office.[12]

The *Time* article has been overtaken by events:

> Chiat Day, an advertising agency has replaced offices and filing cabinets by couches. Given their druthers, almost half the staff telecommutes from home or the road via pager, cellular phone, computer and modem. There is a 'concierge desk' where employees can book an office, pick up a laptop computer and portable telephone. One vestige of old times is conference rooms. Paper "has all but vanished" in favor of messages on personal computer screens.[13] The homey collection of photos, plants and souvenirs once found on and around private desks and offices are gone; personal effects are stowed in employee lockers.[14]

As Jeremy Rifkin writing in *The Nation* states:

> Automated technologies have been reducing the need for human labor in every manufacturing category. Now, however, the service sector is also beginning to automate: In the banking, insurance and wholesale and retail sectors, companies are eliminating layer after layer of management and infrastructure, replacing the traditional corporate pyramid and mass white-collar work forces with small, highly skilled professional work teams, using state-of-the-art software and telecommunications technologies. Even those companies that continue to use large numbers of white-collar workers have changed the conditions of employment, transferring workers from permanent jobs to "just in time" employment, including leased temporary and contingent work, in an effort to reduce wage and benefit packages, cut labor costs and increase profit margins.[15]

All told, at least 80 per cent of all employed persons will probably not be awarded the flexibility in routine and location resulting from advances in telecommunication.

For those in the remaining 20 per cent there will be more flexibility. College professors are one historically autonomous group, predating the computer, a profession famous for flexible hours and independence. Writers and artists supported by their profession also fall in this category. What is new is the increasing numbers of back office staff of computer programmers, researchers, management analysts, financial staffers and marketing personnel who are now working away from central offices or who can and will be doing so in the course of the next decade.

Even at that some of this group are optionals: travel agents, word processors, legal assistants, some engineers and scientists, business entrepreneurs, bill collectors and many others can work out of their home or away from the central office at least part of their time or enough to get in an extra day or two at home or to avoid peak hour travel. In short, it is not either/or, professors vs sanitation workers, traveling salesmen vs the assembly line.

Confronting a New World

The predicted world of rich, knowledgeable workers and poor service workers will have many exceptions: There will still be affluent plumbers, and well-paid specialists in foreign car repair and restoration of old masters. One question is whether underpaid, disaffected, low-skill workers will become a breeding ground for civil disorder. We know that the marginals, the dysfunctional underclass, is the source of serious social ills. The question is whether the equivalent of the displaced bank teller, the army of minimum wage retail order-takers and produce baggers, the ex-draftsmen, telephone operators, stenographers, laborers and others scrambling for a bare living will become a legion of resentful *lumpen*, open to conspiracy theories and xenophobia.

Corporate Teleworking: From Privilege to Mandatory Requirement

One stimulus to telework is the Clean Air Act As of November 1994 in thirteen highly populated regions firms with 100 plus employees at one site must submit and implement plans for reducing peak hour commuter traffic. However, the real thrust comes from economic, not environmental or personal gratification incentives. The key is reduced costs for office and parking space as more employees are shifted out of the head-office. Plus, there is the promise of increased, easily monitored productivity, by cutting back on wasted commuting time and office distractions.

It is clear that the numbers of volunteers is being overtaken by conscripts; employees ordered to work at home or at an off-site center. The trend toward mandatory teleworking picked up steam in the mid-1990s.

Business Week estimates that 83 per cent of U.S. companies are now embracing alternative office strategies. The American office is evolving rapidly in two directions. The first is reorganizing workspace for those employees "who must still work in offices" and the second is "shoving everyone else out the door."[16]

To critics who fear isolation of employees and loss of control by management, there are three responses:

First, measurements of output and productivity are often easier to assess when the work goes over the computer. Pretend-work is harder to generate in the home based office than in headquarters with the froth of meetings and memos satirized by Scott Adam's *Dilbert*. One of the reasons that corporations are so readily taking to the practice is the belated discovery that its usually easier to keep tabs on telecommuters than on office staff.

Second, the loss of personal contact is exaggerated. Serendipitous encounters around the water cooler and the coffee machine are still operative. Why? Most telecommuters do not stay away all the time. Frequent office lunches, morning or afternoon meetings still take place. Remember, telecommuters are classified as such if they work outside the head-office one day or more a week. Most are at the central office at least one or two days a week.

Third, as telecommunications bandwidths and process costs decline, we will see more vido-conferencing, virtual meeting places and very soon, two way visuals. A decade hence we may have three dimensional holograms which offer a reasonable facsimile of face-to-face encounters.

It would seem that telecommuting would be the medium of choice for its employees at the big firms like IBM, AT&T and other giants with their sophisticated management style in general and cost containment interests in particular. In sheer numbers, however, the ranks of at home workers who are telecommuting employees are far overshadowed by entrepreneurs. What has happened is this: small businesses that once relied on mail and telephone have tapped into the potential of telecommunications technology that provides speed and flexibility. Telecommunications technology gives added impetus to a decades-long trend in the growth of the self-employed.

The *May 1991 Current Population/Survey* reported approximately 20 million non-farm employees working at home as part of their primary job, i.e., about one worker in six.

Most home-based workers (24.3 percent) are in marketing and sales, 14.9 percent in contracting, and 13.2 percent are in mechanical and transportation, collectively accounting for more than half of all the respondents in the study. Some 12 percent of the occupations are in the services, which include home child care; elder care; processing vegetables for McDonald's; crafts and a variety of endeavors related to the arts, such as composing, film producing, graphic designing, and creating greeting cards; and professional work in education, finance, government, health, law, religion and science. Less than one percent of the job titles are computer related. On average, home workers are 35 or older. More are married than single, business owners than wage earners, and most work full time. On a mean scale, they have received 13.9 years of schooling, more than is typical of workers in the traditional workplace.[17]

About half the home-workers were full-timers, a quarter were in marketing and sales, a sixth in contracting and an eighth were truck drivers or in other transportation. Only 6 percent were employed in clerical and administrative support and about one in eight was a craftsman or artisan, most of them presumably part-timers. Interestingly, the average home-based business owner earns about a third more than the average wage earner and there are other advantages.

At-home work can increase profits by decreasing operating expenses, and eliminates hours spent on the road and away from home. Federal tax regulations allow self-employed home-based workers to shelter profits under reinvestment, retirement, and other plans. New technology, a growing pool of temporary workers, and new home-based support services that provide everything from document archiving to conference space are making home-based work more "professional," more affordable, and more appealing.[18]

One of the unexpected findings of the mid-1990s surveys is the high percentage of city workers who work out of their homes; teleworking is not just for suburbanites.

PERCENT WHO LIVE AND WORK AT HOME

Downtown San Diego, CA	9.4
Midtown Manhattan, NY	7.1
Downtown Manhattan, NY	7.0
Century City/Beverly Hills, CA	6.4
Bethesda/Chevy Chase, MD	6.1
Austin Downtown, TX	6.0
Berkeley, CA	5.5
Greenwich, CT	5.5
Calabasas/101 Freeway, CA	5.2
Santa Monica, CA	5.2

Source: *Home Office Computing*, November. 1996,24.

Surveys reveal another interesting finding: The data appear to demonstrate a strong gender difference: Two-thirds of home businesses are owned by women. The largest single home business category is business support services followed by desk top publishing, consulting and retail sales from home. All told, these four sectors account for almost two-thirds of the total.[19]

The seers who follow the great American tradition of looking optimistically into the future after each new invention have greeted telecommunication as the pathway to all sorts of joy to come. Following the yellow brick road to Oz will find the elderly, the frail and the disabled granted the gift of mobility and employability. Persons trapped at home as caregivers for the elderly or young children will be able to hold part-time or full-time jobs. Everyone who can pay the modest entrance fee for the requisite equipment can enter a fascinating world of entertainment, information and constant contact with friends, relatives, colleagues and fellow members of interest groups.

In the late 1990s, the pattern is pretty much a matter of current computer-phone-fax technology. The publication, *Home Business News Report,* sees elaborate video conferencing rooms giving way to desktop conferencing but does not speculate on the significance of universal visual communication.[20]

In short, up to the early 1990s telecommuting was seen as a kind of rare privilege granted to self-disciplined employees who could make a convincing case for leaving an empty desk at the office. Increasingly, it is the employer who insists and the employee who must accede. For example, aggressive telecommuting programs at Pacific Bell resulted in one-quarter of the 19,000 workers telecommuting at least one day a week, and similar efforts were under way at Compaq, Perkin-Elmer and Hewlett Packard. These are all leading-edge firms based in California. When the Los Angeles earthquake added to chronic peak hour traffic congestion, it spurred the trend toward structured programs aimed at cutting corporate operating costs. Companies have decided to reduce office costs, parking spaces, motor pools and lost travel time by ordering employees--usually beginning with the sales department--to work away from the central office. It is noteworthy that Denver's travel reduction program, TRP 2000, aimed at "squeezing every nickel" teaches management teams to save money with innovative work strategies. The city and county of Denver's three day certificate program advertises that "its graduates include executives from life insurance, airlines, utilities and other firms."[21] Managers should be "work product-oriented." Benefits include retraining valued employees, reductions in office space, "happier, lower stressed employees in terms of medical costs, sick days, absenteeism and burnout...."[22] The brochure for the program asserts,"Telecommuting is a win-win-win concept...." It further claims that "telecommuters are happier and more productive as are their families, their employers benefit from economic and efficiency gains and their community infrastructure is helped by their presence."[23]

Xerox has its virtual sales office program for sales reps in its Southwest sales and marketing territories. Xerox sales reps have been given the tools that largely eliminate the need to come into the office. Instead district offices will act as business hubs used by roving employees.[24]

One basic financial incentive is available to the draftees. While the volunteer telecommuters willingly assumed the costs of equipment for home offices, the employers pay the expenses for the new breed of conscripts. This can include office furniture, computer, printer, software and fax and one or two extra phone lines. Why two? One is for checking e-mail, faxes and connecting to home office computers and the other is for business voice calls. Hewlett Packard supplies telecommuters with duplicates of home office UNIX workstations. The cost: approximately $4,000 per year per employee. The payoff, more productivity.[25]

Perkin-Elmer offered substantial incentives for its sales and service engineers when it consolidated its outlying sales offices into seven.

Each employee received a laptop computer with modem, two telephone lines and a $1,000 furniture allowance.

Hewlett-Packard offers the telecommuter alternatives to an increasing number of "non-field" staff such as design engineers who need blocks of uninterrupted time transcending standard office hours.

Given huge cost savings from reduction in-office space needs, large firms have begun to leap into telecommunication technology, a trend that is beginning to have an impact on federal employees (state and local government are a decade behind).

In short, the world of work is changing fast; choices about where to live and work are increasing by quantum leaps, particularly as technology-oriented boomers move up to senior management status. Mokhtarian suggests that the 8.2 million telecommuting employees in 1995 could be substantially increased were it not for manager reluctance. In her view, 16 percent of the US labor force that could telecommute shrinks to only 6.1 percent partly because of this employer resistance. Given the one-two days a week typical telecommuting pattern, this translates into a minor 1.5 per cent reduction in highway travel. It seems likely when manager resistance is replaced by manager insistence, the local impact may be substantial and the move out of expensive downtown headquarters office space will accelerate.[26]

By the late-1990's services can be provided by countries thousands of miles apart as the transmission of information will permit and indeed foster enormous changes. Proximity is no longer a given. Possibly, the pressure for opportunity-driven migration from third world nations will diminish as professionals can remain in place while working for a far off first world corporation. On the other side of the world, skilled professionals in advanced nations will find themselves under siege by lower-paid third-world competitors. And in rural and remote areas in the first world, telecommunications may open up opportunities for local professionals and migrating workers from big cities.[27] The prescription cannot be made generic: The key issue will be the relative attractiveness of local amenities. A staggering number of communities offer very little for people with wide residential options. *The Economist* foresees a world where the "death of distance" will mean that:

> ...any activity that relies on a screen or a telephone can be carried out anywhere in the world. Services as diverse as designing an engine, monitoring a security camera, selling insurance or running a

secretarial paging service will become as easily exportable as car parts or refrigerators.[28]

Among the "glimpses" of the world of the future, we see India attracting back-office work from Swiss Air and British Airways as well as computerized monitoring for air conditioning, lighting, and lifts elevators in Singapore, Malaysia, Sri Lanka and Taiwan.[29]

Nonetheless most people have no choice. They have the kind of job that makes it impossible or unlikely to avoid the traditional nine to five routine. It's not just assembly workers who have to be there: Nurses, hospital attendants, cashiers, most retail sales clerks, janitors, waiters and cooks, gardeners, policemen and firemen, doctors, carpenters and teachers, mechanics, farmers, miners, fishermen, and receptionists have to be there in person at regular hours.

Telecommunications and the New Urbanism

The telecommunications revolution links economic activities that don't have to be in physical proximity. They offer more freedom to choose where people work, live and go for recreation. The big question facing U.S. urban areas in the next generation is where will people choose?

We must also take into account that many other factors are at work. For example, retail and wholesale trade may be considerably changed via home shopping but it is likely that the advent of the big boxes such as Walmart, has had more effects on retail trade than sales via computer or television.

There can be no quarrel with Graham and Marvin's depiction of urban areas as much more than aggregations of structures bound by traditional linkages. In their view:

> ...contemporary cities are not just dense physical agglomerations of buildings, the crossroads of transportation networks, or the main centres of economic, social and cultural life. The roles of cities as electronic hubs for telecommunications and telematics networks also need to be considered. Urban areas are the dominant centres of demand for telecommunications and the nerve centres of the electronic grids that radiate from them. In fact, there tends to be a strong and synergistic connection between cities and these new infrastructure networks. Cities--the great physical artifacts built up

communications whose traffic floods across global telecommunications networks--the largest technological systems ever devised by humans.[30]

What is important are the next questions. Which cities fit best into this new world? Which areas outside central cores will come aboard and which will lag? In short, in an era of choices, who will be chosen and who will be rejected?

We can agree with Graham and Marvin that in this emerging urban world there is an

> ...inherent logic of polarization, which seems to be locked into current processes of economic and social development in cities. This polarization is both reflected in, and supported and reinforced by, the development of electronic spaces. Fewer city economies seem set to do well; patterns of economic health become more starkly uneven at all spatial scales; and processes of change seem to reinforce the privilege and power of social elites while marginalising, excluding and controlling larger and larger proportions of the population of cities.[31]

John Keegan sees the decline of older US cities as inevitable, victims of emerging technologies that shifted the focus of development outward. In his view:

> The old cities have lost their hearts because they were built by people who thought at a foot's pace, journeyed by horse. The vastness of America, for all the heroism of early journeys made by foot or horse into its unexplored interior, demanded other means of motion, the locomotive, the motor car, the airplane, means of devouring space, not of submitting to it. It is the space that surrounds American cities, the interminable distances between them, that have done for small streets and town squares, felled the shade trees, left the porticoed churches standing amid desolation, driven freight yards and interchanges and airport expressways into the order that once was. It could not have been otherwise. Once Americans decided to command their continent from coast to coast, all three thousand miles of it, to have no internal frontiers, to spend a common currency, to obey, often not to obey, a uniform code of law, to recognize a single government, to be one people, the life of the small city, the shape of the pedestrian neighbourhood, was

doomed. Traveling America confronted settled America and traveling America triumphed.[32]

Keegan overstates his case, particularly when he expresses admiration for a number of old cities that have retained their charm and vitality, cities like Charleston and small towns like Annapolis and Stonington, Connecticut. But the thrust of his argument is difficult to counter.

Graham and Marvin cite the response of a prominent British urbanist, Richard Rodgers, to the crucial issue of the survival of central cities in an era of social and geographic polarization.

> What, he asked, would be the the fate for Britain's cities if a new set of urban ideals, and the mechanisms to achieve them, were not built up to address the growing sense of urban crisis in Britain? Rodgers' response was stark and simple: 'Blade Runner.' The poor will be ghettoised in their estates, walled in by police and by the barriers of unemployment. The rich will be in their ghettos too, electronically fortified. Everyone will be separated in his or her own security castle. There will be no society.

> ...much of what we have found in this unprecedentedly broad review does seem to support Richard Rodgers' rather pessimistic outlook. As part of the ongoing economic, social and cultural change surrounding the shift to post-modern urbanism, telematics do seem to be helping to support the emergence of new, more highly polarised social and cultural landscapes in cities. The truly public dimensions of cities where citizens interact and encounter each other in physical space seems threatened. Urban trends seem to be supporting instead a shift towards tightly regulated private and semi-private spaces--both physical and electronic--oriented towards the exclusion of groups and individuals deemed not to belong.[33]

As one cynic puts it, the information highway is the only highway that doesn't go through the ghetto. (This refers to the fact that in-town roads tend to avoid upper income neighborhoods and seem to be magnetically drawn to slum areas.)

We can reasonably suggest that teleworking will accelerate--give added impetus--to trends that were in full operation before the first computer was unpacked from its crate. What can we realistically expect in the next ten to twenty years?

Central Cities and Inner Suburbs

Mostly more bad news: Headquarters office buildings, taxes and jobs have been a major prop for central cities that have lost most of their manufacturing and much of their retail base. We can anticipate more downtown office vacancies along with some attempts, mostly frustrated, to adapt old buildings for residential use. Furthermore, there will be more losses of jobs. The problem is not engineering: we can clean up the toxic brownfields, build mass transit, climate control mega environments. Engineering is easy, albeit expensive. Poor quality governments, bad schools, high crime rates and in general dysfunctional populations have so far proved much tougher obstacles. Even the happy news that crime rates have fallen sharply in most central cities since the early 1990s is not likely to alter the consensus that cities are dangerous places unless the crime rate goes lower and stays low.

Inflows of empty nesters and adventurous youngsters help but the chief hope seems to be partial population replacement through large-scale foreign immigration. Hardworking, law-abiding, education-oriented, property-conscious immigrants have been the salvation of many rundown urban neighborhoods. A welcome mat for industrious immigrants rather than restrictive legislation may hold some of the answers to core city problems.

Suburbs

Within the diverse collection of communities in US suburbs new technology is likely to accentuate existing patterns. The poor inner suburbs which have even less going for them than the adjacent central cities will continue to remain impoverished. In contrast, affluent suburbs with high quality schools and other attractions will continue in the "winners circle," much sought after by the well-off and subject to departure by the downwardly mobile downsized and unlucky.

We can also expect that despite the relatively greater suburban safety where crime rates are 80-90 per cent lower than in nearby central cities, even moderate income suburbs will experience the "gated community" phenomenon which is spreading throughout the nation. The crime sensitive will pay a premium for the perception of increased safety.

So far as travel is concerned, we can expect increasing resort to teleworking to diminish commuting time, cost and effort: some of it in the home, probably most of it in small suburban centers. A surge in telecenter construction is on the near horizon.

In support of the axiom that to those that have more is given, upscale suburbs (and their exurban clones) have a leg up in the amenities wars. They are out in front in the competition for the home based businesses and local centers that will function as substantial industrial incubators in the future. To exploit their advantages requires only minor adjustments to zoning regulations in single family zones and restraint in imposing licensing fees.

Exurbs

The tide of population and economic growth has been flowing toward communities twenty to thirty miles out or even farther in the case of larger metropolitan aggregations. We can foresee more of it, faster and on a substantially larger scale.

One problem with many distant suburbs is their tendency to sell themselves short. Rather than recognizing that they can now afford to be increasingly choosy about design standards and potential incoming firms, they are inclined to roll over for almost any developer who can get bank financing. The result is likely to be a kind of secession: many upscale newcomers will locate in separate, gated developments.

There are several problems that confront relocators to outlying fringe areas. One is the mediocre quality of local schools. The response may be charter schools, home schooling or willingness to pay high tuition costs at private schools. A troublesome problem may be crime. Move way out to an area with pockets of primitive natives and you may find yourself confronting the road company of *Deliverance*: alcoholism, vandalism and violence. (Gated communities may be one response to this menace.)

Another challenge: medical care. No problems arise if the individual and family stay healthy, but a complex illness or injury may require more than the local MDs and clinic can offer. Not an easy one to solve although in some instances helicopter evacuation to a university teaching hospital can be a nick-of-time solution.

Recreation Areas

Over the past two generations population in many resort areas, particularly coastal resorts, has been increasing at two or three times the national average. The technologies that made it possible were advances in transportation, especially air travel, telephone and the critical climate control device, air conditioning. Teleworking adds to economic possibilities in these areas, as the increasing number of persons in bathing suits attached to their cellular phones attests. In general, the geographic movement (which includes retirees who earn North and spend South) has been from the icebox to the balmy. Increasingly people live year round where they used to play a small part of the year. Florida is an outstanding example. In an age of talent-oriented growth rather than resource based economic development, jobs in industry, commerce and services have followed the people. But there are flaws in the ointment: Many resort areas are plagued by natural disasters: hurricanes, floods, earthquakes, mud slides and fires. Two other problems may be mentioned. Like many of the exurbs, many resort areas have mediocre schools, so-so local government and often, chancy medical quality, especially in relatively remote playlands.

College Towns

If there is one kind of community that has it all, this is it. Indeed places like Princeton, Chapel Hill, Palo Alto and other centers of charm and research have become targets for economic location, particularly for high tech industries and services. College towns usually have good schools, good government and good doctors along with low-crime rates and an appealing intellectual-cultural environment. Not surprisingly, housing prices have gone through the roof in some of the most attractive college towns, but fortunately there are less expensive communities adjoining or close to the ivy.

Social Consequences: Polarization/Fragmentation

The broadening of geographical choices for an increasing proportion of the population can have serious negative consequences. One of these is the reduction in physical proximity between social classes. It is becoming possible to pursue a career in business, professional or technical occupations and live one's life without having any but the most fleeting contact with poor people. And the poorest may stay poor longer. An underclass lacking literacy and numeracy may be counted out of the mainstream when the entrance fee is raised to include teleworking skills.

Finally, there is on the near horizon the prospect of two technological advances. The first is universal two-way video communication. This is likely to reduce further the need for personal, face-to-face contact. But the giant leap toward depersonalization may come with virtual reality: widespread use of headsets and gloves and down the road, body suits may have enormous impact. We can foresee a huge increase in couch-potato dropouts, heroes in private universes where rejection never happens and victory in love and war is a certainty. Experience with Dungeons and Dragons, that frightening teenage addiction, is only a foretaste of things to come. One need hardly say that therapy counseling will be one of the growth industries of the next century.

Conclusions

This analysis of things happening and things to come is a description, not a prescription. It suggests that remedial corrective action to reverse those trends that many of us find alarming is extremely difficult. We may have to look for partial solutions in past history. I suggest, for example, that if there is a partial answer to this dawning age of social fragmentation, it may come in the form of some type of compulsory universal service to guarantee the kind of interclass contact that was one of the beneficial byproducts of military conscription.

1. Office of Technology Assessment, *The Technology Reshaping of Metropolitan America*, (Washington, D.C.: US Government Printing Office, 1995), 165-166.
2. Ibid, 17.
3. Ibid,171.
4. Jack Nilles, ,"Telecommuting Forecasts," *The Technology Reshaping of Metropolitan America*, Office of Technology Assessment, (Washington, D.C.: US Government Printing Office, 1995),170.
5. Richard Nelson Bolles, *The 1996 What Color is Your Parachute?* (Berkeley:Ten Speed Press, 1996),115.
6. Karen Kaplan, "For Workers, Telecommuting Hits Home," *Los Angeles Times*, July 29, 1996, D7.
7. Jon Nordheimer, "You Work at Home: Does the Town Board Care?, *The New York Times*, July 14, 1996, Section 3, 1
8. Dr. Charles Grantham, President of the Institute for Distributed Work, quoted in Susan J. Wells, "For Stay Home Workers, Speed Bumps on the Telecommute," *The New York Times*, August.17, 1997, 14.
9. Ibid, Section 3,4.
10. Hedrick Smith, *Rethinking America*,(New York: Random House, 1995), 210.
11.*Time*,"Special Issue: Welcome to Cyberspace," Spring 1995, 37.
12. Ibid, 38
13. Time, Ibid
14. Ibid, .38-39
15. Jeremy Rifkin, "Civil Society in the Information Age," *The Nation*, February 26, 1996.
16. "The New Workplace," *Business Week*, April 29, 1996, 109.
17. Ibid, 16.
18. Ibid, 18.
19. *Home Business News Report*, PT Corp, 1996.
20. Ibid, 117.
21.City and County of Denver, TRP 2000
22. Ibid.
23. Ibid.
24. Deborah Lewis, "Telecommuting: Round Two--Voluntary No More," *Forbes* ASAP, October 9, 1995, 133-134.
25. Ibid, 138.
26. Patricia L. Mokhtarian, "A Synthetic Approach to Estimating the Impacts of Telecommuting on Travel," paper prepared for the TMIP Conference, Williamsburg, VA, October 27-30, 1990.
27. "The Revolution Begins, At Last," *The Economist*, Vol. 336, No. 7934, September.30, 1995, 15, 16.
28. Ibid., 227.

2 9. Ibid., 27-28.
3 0. Stephen Graham and Simon Marvin, *Telecommunications and the City*, (New York: Routledge, 1996), .3.
3 1. Graham and Marvin, op.cit., 378-379.
3 2. John Keegan, *Fields of Battle*, (New York: Knopf, 1996) 16.
3 3. Graham and Marvin, op.cit., 234-235.

for city poor rationalized as "tough love" to force the welfare population into the labor force.[2]

The miseries that afflict central cities have faded into a minor issue; first, among conservatives who mostly lived and voted in suburban communities and since the 1970s, the bulk of the liberal-leaning electorate has also become suburban with the result that urban problems which loomed so large in the political agendas of the 1960s and 1970s have been pushed to the back burner. The decline in the proportion of U.S. population living in cities from one-third in 1960 to less than a quarter by the late 1990s is reflected in a loss of political clout. By the 1960s the Republicans were regularly winning elections while losing the cities, and by the 1990s phrases like "urban crisis" had fallen off the political agenda. Indeed, by the mid-1990s, a bipartisan consensus had emerged to the effect that the problems afflicting central city slum populations were deep and resistant to government intervention because they are in some measure, self-inflicted.

Are There Jobs There?

Writing in the late 1960s, Anthony Downs asserted that the millions of jobs still located in central cities could provide potential openings for central city poor simply by virtue of normal job turnover.[3] He believed that the key to progress for the most seriously afflicted city population (ghetto blacks), is for employees and other workers to cease racial discrimination in their hiring and promotion practices. More recent research supports Downs by underscoring the job potential of the central city.

> The most troubling aspect of in commuting is not the stereotypic professional but the impact of clerical and service workers coming into the central city to compete successfully for jobs that should be well within the horizons of the central city poor. A depressing statistic capsulizes the problem. Stuart Eizenstat, Chief Domestic Policy Advisor to President Carter, estimated in 1980 that "about 41 of every 100 black teenagers graduating from high school are not literate and therefore have serious difficulties competing for jobs.[4]

Employer preference for immigrants represents another barrier to employment for the native central city poor. As Wilson sees it "research has consistently shown that migrants who leave a poorer economy for a more developed economy in the hope of improving their standard of living, tend to accept, willingly, the kinds of employment that indigenous workers detest or come to reject."[5]

A Boston study disclosed that, in the late 1970s, one-seventh of the 150,000 jobs in the city's expanding private service sector were classified as "low-grade" jobs.[6] Thus, it appears that the problem for central city residents is not so much that low-skilled employment is scarce but that so many of the available jobs are held by suburban residents or new immigrants. Incommuters from the suburbs included sizable numbers of semiskilled and relatively low-skilled clerks, waiters, custodial staff, and security guards--positions for which many central city residents could qualify were it not for employer preference in hiring and retaining suburban residents.

William Julius Wilson offers a devastating catalog of reasons for rejecting central city ghetto job applicants:
- They talk street talk, not standard English,
- they can't read, spell, write or do basic arithmetic,
- they are poorly groomed,
- they cannot adapt to work disciplines including punctuality, showing up regularly and taking direction and correction.

These views, Wilson reports, are shared by African-American employers whose attitudes do not differ significantly from white employers.[7]

Brooklyn's Red Hook area is a rarity among inner city neighborhoods. It has a large and growing number of well-paying, unskilled, blue-collar jobs, but local residents don't get many of them because employers prefer immigrants over the African-American and Puerto Ricans who live in nearby slums. What is more, they almost always fill new positions using private referrals, not advertisements. So even those in the community who might be qualified for jobs rarely know when they're available.[8]

The tendency is to write-off locals as hopeless cases. As one employer put it.

"Look, I'm a liberally minded guy, and in the past I made a real attempt to hire disadvantaged people. But the problem is that I need someone to come to work every day and the family background of these people is so poor that they've lost the work ethic...Do you want someone from that legacy, or would you rather bring in a hard-working skilled person from the [former] Soviet Union, who just got here, who's had an education, and who doesn't know from the drug business?"[9]

The principal worries about hiring locals seem to be potential criminal behavior and lack of work discipline. There is a persistent view among local employers that

> ...immigrants have a better work ethic than native born blacks and Puerto Ricans. The result is that there are many blacks working on the waterfront, but often they are West Indians from other parts of Brooklyn. Lots of Hispanics have jobs there as well, but they tend to be recent immigrants from Mexico or other parts of Latin America.[10]

These comments from a few employers in one section of Brooklyn might be partly discounted as unrepresentative, local and unscientific, were it not corroborated by a substantial body of research.

While Wilson sees the decline in accessible entry level reasonable pay jobs for central city males as a major cause for high rates of unemployment and dysfunctions, much of the problem seems a matter of employer choices from the pool of available applicants rather than a dearth of jobs. As Wilson sees it, a survival culture among black teenagers which teaches them to evade street confrontation by avoiding eye contact and looking tough is a poor foundation for prospective job seekers.

The growth in central city jobs that has taken place in recent years has not been entirely confined to highly skilled white-collar jobs, as had been feared. It is true that the number of entry-level occupations, (i.e., requiring less than high school graduation) has declined, while the number of knowledge intensive jobs has risen. Nevertheless, there remain millions of central city jobs for the less educated in restaurants and hotels, in transporation, and other sectors and occupations where immigrants have found a foothold. It should be remembered that numerous entry level jobs are made available through turnover--firings, resignations and retirements. Stability or even shrinkage in a sector does not mean that there are few openings for fresh manpower. In fact, there are sufficient numbers of semi-skilled and lower-skilled jobs to accommodate most of the unemployed in central cities, if, in fact, the jobs were held by them rather than by in commuters or immigrants. In short, some of the standard explanations for differences in income levels and unemployment rates between central cities and suburbs don't stand up under careful scrutiny.[11]

The Labor Force in The Slums.

Both conservatives and liberals agree that there are large numbers of potential workers in the slums. Liberals believe that the high unemployment rates in slum areas (usually double the city averages) are due primarily to a shortage of jobs for low-skilled people. In contrast, conservatives often suspect that most joblessness is tied to laziness, malingering, and semi-legal or outright illegal hustling. An examination of the 1990 Census tract data for Baltimore suggests that the problem is much more complex. Comparing census figures for the 10 per cent highest income tracts with the 10 per cent at the lowest end of the scale, each with about 75,000 people, it is clear that low school test scores, out of wedlock births, low birthweight, poverty, disabilities and crime are found at higher rates in the low income tracts.

For example:
- School achievement test scores in 1990-91 were substantially lower in the poverty census tracts (than in) in the highest income areas.
- Out of wedlock births were almost 90 per cent in the poverty area vs 30 per cent in the high income area.
- A seventh of the babies in the low income area were classed as "low birth weight," double the percentage in the high income area.
- Almost 70 per cent of the children in the low income tracts were living in poverty in 1989 vs 5 per cent in the upper- income area.
- The percentage of persons with reported disabilities was over twice as high in the poverty areas as in the upper- income tracts.
- There was a 20 percentage gap in labor force participation between the poor tracts and the upper- income tracts.
- In 1991, the total number of reported crimes in the low income area represented a figure of about one seventh of the total population, double the figure in the upper- income area.[12]

Given these factors of low labor force participation rates, high rates of crime and health and other problems, short-term, employment-focused solutions are unrealistic. Moreover, present trends in out of wedlock births, low birthweight babies, low school achievement, and other indices of problems in the making, present us with an even more troublesome future.

Reversing the Tide

Ever since the prospects for central cities began to dim, three methods of effecting a turnaround were set in motion. (Interestingly, it was a fourth, mostly unplanned, avenue of change that had the most impact.)

The arithmetic seemed simple in the 1960s: Factory jobs were leaving the cities. These places had an infrastructure and labor force accustomed to factory work so why not create industrial parks, float revenue bonds and give tax concessions for industry? The result? After thirty-odd years of turning the spigot on, saving some local firms and attracting new ones, the net losses still have been catastrophic: new jobs couldn't fill the vacuum left by heavy job losses. A case in point: Philadelphia had almost 236,000 manufacturing jobs in 1970. By 1996, the total was under 60,000, a net loss of nearly 75 per cent.[13]

However, despite overall losses, industrial employment trends in central cities reveal a complicated set of cross trends. For example, there has been the discovery and rediscovery that cities can be a lot more effective in efforts aimed at retaining existing firms than in attracting new plants. There have been significant successes in many cities in catering to the needs of local firms with deep local roots. But business conditions can turn the most promising situations sour when the market declines. Outside competition can be too formidable even for hardy local survivors, mergers wipe out local firms or plain bad luck can make further operations impossible. Another path to industrial growth is selecting high tech industries as industrial park targets (biotech in the case of Baltimore) on the premise that high wage growth industries will be the engines to propel the city economy to prosperity. The risk is that as they mature, high tech firms with limited local ties will relocate to the suburbs or elsewhere after the pioneers go public or sell out to a larger firm.

Industrial incubators, the use of old industrial buildings to house fledgling firms is another promising avenue. Baltimore, for example, has over 100 firms with 11,000 employees in its old Canton industrial area.[14] The problem is that while the gains are real, the firms are often fragile, subject to buffeting by market forces over which they have no control or else they remain small niche survivors.

A number of cities have benefited from these unintended and unforeseen consequences of plant shutdowns. Some managerial employees have used their severance pay to start their own successful firms. Examples include post-Boeing Seattle in the early 1970s, and in Rochester, New York after General Dynamics shut its electronics plant in 1971.

> Within 15 years, some 17 separate companies in Rochester had sprung up from the ashes of General Dynamics, collectively employing three times as many workers as had been laid off.
>
> ...in the Seattle area after Boeing fired 17,000 workers in the early 1970s...some 600 to 700 new companies (started up) whose origins could be traced to these layoffs. the same thing after NCR's big cutbacks in Dayton in the mid 1970s.[15]

Does this record hold out much hope for other cities? It seems to depend on local entrepreneurship and the particular time and circumstances. There is no evidence that government intervention played a key direct role. A deciding element was personal ties--to friends, to the community. In short, thanks in part to effective civic leadership and to the fact that both Rochester and Seattle are considered nice places to work and live, they came through hard times. The path to a turnaround is increasingly being sought in the market place rather than through government intervention.

As Jon Jeter reports

> ...the nation is rethinking its strategy to repair its broken inner cities. Indianapolis' plan does not focus on the construction of affordable housing, job training programs, massive tax benefits or any of the altruistic efforts fashioned from the urban policy mold of decades past. This city's approach to urban renewal is about capitalism, pure and simple....
>
> While providing the poor with affordable housing and preparing them for the workplace are still crucial elements, cities are increasingly turning to the marketplace as an engine for social change.
>
> In effect, this strategy looks at urban renewal through the eyes of business. Rather than simply bribing companies to locate in a certain spot with multimillion dollar tax cuts, the idea is to convince businesses that slums can actually be profitable. There may be more money in the suburbs, the pitch goes, but there's also a lot more competition. Inner-city neighborhoods, by contrast, tend to have

high population densities and hardly anyone chasing after their money.[16]

Another possible path for city development is the redemption of urban sites contaminated by toxic wastes. An estimated 250,000 brownfields (mildly contaminated industrial sites) lie like scars on America's urban landscape, a legacy of the dramatic shift of industry from inner cities to suburban "greenfields."
The revitalization of brownfields is one way to breathe new life into moribund urban areas across the country.[17] One method of luring new business onto brownfields is to provide tax incentives. Some states like Ohio are encouraging voluntary private action, partly by reducing environmental standards for such areas and providing legal safeguards against future lawsuits. Will it work? Some progress is likely, and scattered, anecdotal reports of successes can be anticipated. But the prime obstacles remain: it will take substantial subsidies to persuade private firms to relocate in decaying areas, particularly when there are competing sites in attractive suburban and exurban environments. As OTA points out, thousands of these contaminated sites are inside cities where industrial employment has declined and this vacant or underused property is barely on the tax rolls. Unfortunately, redeveloping these sites is far from simple because
- it costs a lot. Generations of poisoning and neglect make the cleanup slow and expensive;
- There are serious and complicated legal problems of liability relating to potential health problems;
- Ownership may be hard to disentangle.

In recognition of these obstacles, OTA's analysis suggests a need for caution, particularly because these sites are, after all, located in cities where foreseeable market demand for new industrial sites may range from weak to non-existent.[18]

It is possible that the brownfield problem may be overstated. For example, in a 1996 survey of contaminated sites in Baltimore, Howland discovered that many sizable brownfield sites have been successfully marketed--after substantial price discounting.[19]

As the case in Baltimore, in a number of cities the market for brownfield sites--at least those that are moderately contaminated--seems to have taken a turn for the better. The key to improvement is relaxation of federal guidelines and a scent of profit by the private sector.

Beginning in 1994, the Federal Environmental Protection Agency quietly relaxed its formulas for the brownfields. A retail use, for example, would require a less rigorous, and less expensive, clean-up than a residential use. That change created a new industry, which arose to clean up and resell the properties.

At the same time, the cataloging of brownfield sites became a function of local government, which can respond to the needs of business faster than the Federal bureaucracy. And perhaps most crucially, investors who were looking for returns greater than the 12 percent to 15 percent typical in mainstream real estate saw urban values appreciating as the crime rate declined.

"On a brownfield site, you can make 20 percent to 30 percent, if you do the numbers right," said Christopher J. Daggett, president of Chadwick Partners, an Edison, N.J., company formed 19 months ago to get into the brownfields business."[20]

In passing, it should be noted that many cities have a wide variety of sites to offer. They lack only buyers. Economists make a careful distinction between demand and effective demand. The former connotes need. The second is need plus the requisite money to satisfy it. Cities are full of people who need jobs. They have sites and buildings that can house office workers and manufacturing plants, and vacant areas suitable for new housing. What's missing are sufficient takers for what they have on the market. Overall, with notable exceptions, Boston; San Francisco; Portland, Oregon; and some others, central cities have thousands of parcels of vacant land on the site of abandoned housing or decayed commercial or industrial plants but there are no takers. Alexander Garvin is right on target when he identifies market demand as the key ingredient in the many successful projects he reviewed.[21]

How much vacant land is there? Take Baltimore as an example. Baltimore has 14,500 parcels, 8,000 "occupied" by vacant housing and 6,500 simply vacant. Some neighborhood blocks are half vacant, victims of a population decline that has an infrastructure in place for a city almost a third greater than its present size.

Redraw the Boundaries

In the nineteenth century central cities regularly annexed their suburbs. In the twentieth century this practice is in evidence only in parts of the South and West. Troubled cities in the North and East could not, for political reasons, incorporate the affluent residents, good jobs and tax base in their suburbs. As David Rush reminds us, since 1950 half of America's central cities more than doubled their territory by annexing new suburbs and another twenty cities (notably Nashville and Jacksonville) in effect annexed their counties. Not one of these near-regional cities has less than an "A" credit rating.[22] Unfortunately, boundaries are effectively frozen for many of the most troubled cities. In the absence of growing suburbs, these cities have state and federal subsidies to help out with their welfare, crime, environmental, educational and other problems. Indeed, if the political will were there, the much-lamented fiscal stress afflicting central cities could be alleviated by a stroke of the pen. The difficulty is that other jurisdictions have problems of their own. They don't feel all that affluent or charitable. They view central cities as chronically misgoverned, scandal-ridden cesspools with much of their population as hopeless losers, undeserving of more outside help, and in the event of merger, a threat to good government and good schools.

The preferred alternative to outright annexation is "functional metropolitanism," assigning specific functions like mass transit, parkway and water/sewerage systems to special metropolitan authorities while leaving municipal governance, schools and police forces in local hands. A second approach is neighborhood governance, giving city neighborhoods substantial responsibility for public schools and other services. This alternative was popular in the 1960s when it was seen as "empowerment" a method of stimulating participation of the poor in municipal governments that had too long comfortably co-existed with urban poverty. The results? Promising in some areas, in others an avenue to discord between City Hall and neighborhood groups, allegations of nepotism and corruption, little evidence of higher school test scores.

The other side of neighborhood governance is attempts at reverse triage: This includes attempts at suburbanizing middle- and upper-income neighborhoods. As examples we have Staten Island's occasional threats to secede from New York City and recurring proposals to transform Washington, D.C.'s middle- and upper- income neighborhoods into Maryland communities like Takoma Park, leaving slum areas and downtown as central cities. These proposals seem to have gone nowhere.

Another Answer: Move 'em Out.

During the 1960s and 1970s much academic ink and huge legal fees were expended on the "gild the ghetto" versus "disperse the ghetto" controversy. Economic development and Model Cities programs were based on the premise that cities needed to create jobs near where the poor lived-- "gild the ghetto." The alternative, "dispersing the ghetto" was based on the opposite premise; i.e., the critical mass of dysfunctional inner city residents could best be saved by stimulating the movement of the central city poor to the suburbs on the basis of "fair shares" with each suburban community receiving a fraction of the central city poor.[23]

The positive results of both approaches have been modest. As noted above, industrial jobs have departed the cities in large numbers and after over a decade of spirited litigation in New Jersey and other states, few ghetto poor have been helped to move to suburban jurisdictions. Clearly one alternative to the community turnaround approach would have been the dispersal of as many residents as possible away from dysfunctional areas into communities where there is access to better schools, safer neighborhoods and better jobs. This was not to be. While strenuous efforts at dispersal have been attempted in New Jersey and elsewhere most of this litigation aimed at breaking down zoning barriers has not resulted in much opportunity for the hard-core poor. HUD Secretary Cisneros tried in the early 1990s to launch "Moving to Opportunity," using housing certificates to allow individual families to relocate on their own but this initiative died in its tracks. In the political climate of the 1980s and 1990s suburban dispersion through government action proved controversial and ill-timed.

Over the years, a handful of dispersion programs, most notably, the Gautreaux program in Chicago, have been successful, albeit on a very small scale. Costs are low, children of relocating families are more likely to graduate from high school, but there is no change in employment levels. Job benefits are most likely to be enjoyed by the next generation, not relocating adults. Most important perhaps; the scale is very, very small: Only 12,000 participating families nationwide.

As Jacob Weisberg explains, the Gautreaux Program worked because a few thousand, individual black families headed by single mothers, were carefully screened and filtered into white suburbs. It also worked because it was tiny, stealthy and carefully designed to exclude the hell-raisers busy wrecking their neighborhoods back in Chicago.[24]

In short, the results are promising but a significant expansion of scale will require a change of national policy which is not likely to be forthcoming in the political climate of the coming decade.

It is likely that in time-honored tradition, the central city poor will have to rely on their own earnings to buy into either upscale city neighborhoods or move out to the suburbs. This means that we will see a continuation of the "two nations" identified by Benjamin Disraeli as early as in mid-nineteenth century Britain and still existing generation after generation in the U.S. and elsewhere in the world. Compared to some previous eras, proximity to the central city poor is now considered hazardous, therefore, teleworking offers increased opportunity to further restrict contact with perceived danger. Not only can you move away to greener communities with better schools, now you can sharply cut back on the amount of time you spend in a central city job.

Keep The Offices?

One of the remaining pillars of central city economies is the location of corporate headquarters and other white collar office employment. For the most part, these offices did not exhibit the same pattern of wholesale flight as manufacturing. In fact, New York City successfully countered a New Jersey raid on the Stock Exchange, and despite defections to the suburbs, office building construction and office jobs remain a major prop to the economy. However, this may change because

- government employment is shrinking;
- managed health care is reducing growth, or in some cases, causing declines in the robust health sector;
- vacancies in corporate office space may reach or surpass 1996 Detroit levels, (i.e.,20 percent) in the next decade as corporate decisions to cut costs will decrease headquarters staff to a minimum. Telecommuting offers a vehicle to achieve that corporate objective rapidly.

Bring Back the Middle-Class?

An alternative to gentrification (renovating rundown areas by attracting affluent people) is a targeted campaign to induce selected suburban demographic groups (e.g., empty nesters and young singles and intellectuals) to return to stable city neighborhoods that offer many of the advantages as the nearby suburban communities where many of them are now settled. This implies giving up efforts to attract families with school age children and focusing instead on population groups tolerant of city ills like crime, grime, inferior public services and higher taxes.[25] The difficulty is that even those relatively insensitive to city drawbacks can be repelled by personal experience with mugging, burglary, spectacular failures at snow removal or prolonged exposure to incompetent bureaucracies. Moreover, many of the special attractions offered by the city are either being duplicated in the suburbs (e.g., good restaurants, meeting places for singles) or are easily available by virtue of a commute not much more inconvenient than the trip downtown from outlying, affluent city neighborhoods. The question is whether it is the cities that will become the niche market for people-with-choices if they find it impossible to effect the dramatic improvements in public schools needed to attract a full range of populations.

Telecommuting/Teleworking and The Downtown Office

It's been said that people go to doctors only when they feel very ill. Cities mount a full court press for new economic development when the jobs fall away and the tax base shrinks. As I have noted, mounting turnaround programs in stricken cities is usually not a wildly successful undertaking. One problem is that the props of the city economy have been eroding. Manufacturing has long been on the decline although until recently white collar jobs have helped to take their place. Now, however, the restructuring of health care, finance, insurance and other service sectors in the 1990s indicates that this big growth sector is likely to level off. Government, a central city employer of first and last resort for much of the working-class, has been on the downswing. A good part of government job growth has been for school teachers and administrative staff. The decline in the central city school age population in most cities has not yet resulted in a similar reduction in employment (thanks mostly to administrative inertia) but the handwriting is on the wall. Employment in downtown offices has leveled off or declined and downtown retail jobs are weakened by the closing of most flagship department stores.

As more white-collar workers telecommute from the suburbs (or telework at home or in a suburb office) there will be fewer lunch time shoppers and retail customers to buoy up remaining businesses. Once the physical commuting habit is broken, the suburbanites discover that the necessary range of services and restaurants is available nearby where the parking is easy (and mostly free) and the crime rate is low.

The option to choose locations where office space is cheaper, crime rates and taxes are lower, traffic and parking present no problems, means that many central cities face even tougher times ahead. Suburban competition has clobbered the central cities in retailing and manufacturing. Services are next. It is no accident that over fifty of the nation's largest cities declined in population in the 1980-1990 decade. To cite one example: By the late 1980s, 60 per cent of commercial office space in metropolitan Baltimore was located in the suburbs. From the late 1960s to the late 1980s Baltimore lost 56,000 manufacturing jobs and 12,000 more in retailing. To compensate, the city gained 34,000 jobs in business services, one job gained for every two jobs lost. However, many of the job gains were captured by suburbanites, while the lost jobs in manufacturing and retailing tended to hit city residents hardest.

As I have noted, Baltimore is the rule not the exception. From 1985 to 1992 U.S. counties that include major cities lost market share in all the key service sectors: legal services, accounting, educational and medical services.[26]

Baltimore is on David Rush's list of thirty-four U.S. central cities that have passed the point of no return by exceeding threshold indicators in at least two of three census categories he identifies. These indicators are a *population loss* of more than 20 per cent from peak levels, a *minority population* of more than 30 per cent and a *city-to-suburb* income ratio of less than 70 per cent. Baltimore's figures: A 23 drop in population, *minority population* of 60 per cent and the income ratio of 64 per cent made the grade on all counts. Others on Rush's list (big cities only) are Buffalo, St. Louis, Chicago, Philadelphia, Milwaukee, Cleveland and Detroit. The poorest are the two virtually all-black small cities of East St. Louis and Camden, New Jersey.[27]

Naturally, Rush prescribes measures to rescue "hopeless" Baltimore Consolidate the city of Baltimore with its major suburb, Baltimore County to relieve the city of its burden as the region's poorhouse; adopt metropolitan-wide "fair share" housing to disperse many of the poor to the suburbs and engage in effective metropolitan planning and revenue sharing.[28]

The political prospects for such a program? Very dim. Rush's arguments for suburban acquiescence to this kind of metropolitanization are familiar. They range from the practical (stop paying for costly duplication of major facilities), fiscal (suburbs pay for central city miseries in higher state and federal taxes), and economic (future prosperity is endangered when the central city falters).[29]

Rush tends to underplay the racial component. Baltimore's poor are almost all black and its suburbs mostly white. His example of successful, partial integration is Montgomery County adjacent to Washington, D.C. By 1990, thanks in part to a variety of inclusionary housing subsidy programs, the county was 73 per cent white, 12 per cent black, 7 per cent Hispanic and 8 per cent Asian. But close inspection of Rush's data indicates that assisted housing units represented only 3.4 per cent of all Montgomery County housing, leading to the conclusion that most of the non-white newcomers were sufficiently well off to purchase units in the open market. In short, income screening was in effect. The county was most emphatically not the recipient of a large number of hard core poor.[30]

Baltimore's shifting balance in office space in favor of suburban locations is the local manifestation of a national trend. By the early 1990s, 60 per cent of the nation's office space was located in the suburbs and the remaining 40 per cent share in central cities is steadily waning. One reason is the massive layoffs in the Fortune 500 companies that built or rented prestigious downtown offices; by one estimate these cutbacks have opened up 250 million square feet of offices for sublease--if tenants can be found. Put another way, this is the equivalent of fifty Chrysler buildings of five million square feet.

While there are a few skyscrapers in suburban locations most suburban offices are no more than six stories. Clearly one of the apparent casualties of the new era is the skyscraper; i.e., the towering downtown office buildings that define the urban skyline. In mid-1995 only ten buildings of over twenty stories were under construction in the U.S. Over two-fifths of all U.S. office space has been built in the

1985-1995 decade and thanks to optimistic overbuilding, the national downtown office vacancy rate is 16.7 per cent. In some cities it is far higher: Baltimore, 25 per cent; Miami, 27 per cent and Dallas 37 per cent.[31] Telecommuting is a small contributing factor in these vacancies.

Telecommuting didn't create the downtown's problem with office vacancies but it makes them worse. Telecommuting--working from home via phone, fax and PC--used to be the employee's dream. But because it saves central office space, employers also are enthralled. In 1993, for the first time, involuntary telecommuters outnumbered voluntary ones.[32]

Who Leads the Way?

In the late 1960s, the beleaguered Mayor of Newark, New Jersey remarked to the press: "I don't know where America's cities are going, but wherever it is, Newark will get there first." Based on the events of the fifties and sixties, the destination is severe population and job losses, high crime and welfare rates and a troubled downtown. Newark's miseries were indeed a precursor of hard times for dozens of other cities.

Another leading candidate for a sorely affected bellwether is Detroit where there is a "skyscraper graveyard" on Woodward Avenue. Some thirty-plus story buildings are totally or mostly vacant. James Howard Kunstler blames Detroit's miseries on the impact of the automobile:

> The city that spawned the auto age is the place where every thing that could go wrong with a city, did go wrong, in large part because of the car. Until a decade ago, Detroit was the sixth largest city in America and one of the wealthiest industrial cities on the planet. In the last ten years alone, its population shrank by 20 percent. since other nations learned to make better cars more cheaply, the city's auto industry verges on extinction. Anyway, the era of the car-based economy is drawing to a close because we can no longer endure its costs. Detroit, in its present necrotic state, illustrates these costs clearly.
>
> Motoring around Detroit in a rented car, one is not psychologically prepared for the scope of desolation. Beyond the decomposing downtown core of skyscrapers--about the size of ten Manhattan blocks--Detroit is a city of single-family houses that go on and on,

seemingly forever, into a drab gloaming of auto emissions and K Mart signs. The innermost ring of houses is now almost completely destroyed.[33]

Detroit's Renaissance Center was built in 1976 by a consortium of corporate investors led by the Ford Motor Company. Cost: $337 million. It was bought in early 1997 by General Motors for $73 million.[34]

Less architecturally-oriented analysts find multiple culprits for Detroit's devastation--people culture, suburban competition, misgovernment. In any event, Detroit stands as stark picture of a problem plagued city that stands as a warning.

In the eighteenth century, Jonathan Swift published a satirical essay entitled *A Modest Proposal* which suggested that the remedy for Irish poverty and overpopulation was cannibalism-for-export, fattening surplus Irish infants for sale as human suckling pigs. Some analysts believed that Camilo Jose Vergara's proposal that Detroit convert its vacant downtown collection of 1920's skyscrapers into an American Acropolis was another spoof. Vergara proposed that buildings like Hudson's department store (which in 1995 had been vacant for twelve years) and other vacant structures could be turned into a picturesque tourist attraction. The buildings would be allowed to crumble to create a place of silence and repose, one of the world's first major collections of ghostly urban monuments.[35]

Naturally, Vergara--who turned out to be entirely in earnest--aroused almost as big a firestorm as the Poppers' with their proposal to exploit the depopulation of the Plains states by facing the inevitable and returning the arid, depopulating region to a huge "Buffalo Commons."[36] In both cases there were cries of outrage based on previous pioneer sacrifice and endeavor and on the dubious proposition that somehow long term decline could be turned around. In Detroit counter proposals to achieve redemption include adapting eighty-four vacant or underused buildings -- nearly, one fifth of downtown--the new uses ranging from housing, shops and restaurants and entertainment to cheap space for new office tenants.[37]

There are a number of happy exceptions to the overall pattern of troubled central cities. Portland, Oregon, Seattle, San Francisco and New York City are cities where social and economic problems are blemishes rather than intimations of a forthcoming apocalypse.

New York City, thanks to its unique attractions to business--and to new immigrants--is an outstanding, example of unquenchable survival.

Outsiders keep calling it unlivable, but consider one crucial distinction of New York: It's the only major American city that, within its 19th century boundaries, has as many inhabitants today as 50 years ago. The other old cities have all been losing population from their original cores.[38]

In the late 1990s there was good news from the nation's largest city according to Blaine Handey :

With violent crime cut nearly in half, with New York statistically safer than Independence, MO., the palpable sense of dread that had blanketed many parts of New York for decades has largely melted away. Tourism is booming, hotels are full, subway ridership is up and the city's colleges are flooded with out-of-town applicants. Crime is down most steeply in the city's poorest neighborhoods.[39]

One question is whether the downward slide that affects most cities is irreversible. If one looks about for signs of hope there are seers who predict happier days ahead. For example, in an age of corporate downsizing and central staff shrinkage, of 15 and 20 per cent downtown office vacancies there is *The Economist,* suggesting that there is a future, perhaps a bright future for the central office.

A decade or so ago futurologists predicted the arrival of the paperless office. Today they are predicting the death of the office block itself as an ever-increasing number of white-collar workers choose or are told, to telecommute from home. Skepticism is again in order.

In theory, it is possible for many white collar workers to work from home and communicate with each other by telephone, fax, modem, e-mail and Internet. In reality, business colleagues still want and need a central place to meet each other as well as clients, and to exchange ideas.[40]

How valid is this analysis? Clearly, people in a position to choose may be able to retain their central offices if they wish just as a select few have been able to reserve private secretaries, but the reality of the bottom line points to less freedom; i.e., mandatory evictions. If preferences were all that mattered no one would choose to carry on essential dirty jobs. Most employees do not have the option and many white collar workers have to adjust to new conditions--doing all or most of their work away from headquarters offices to earn their paychecks.

There is a second point. We are not presented with a rigid dichotomy: work downtown or work at home. In reality the most popular form of telecommuting may be occasional, off-peak work at headquarters combined with work at home or in small, suburban clusters which offer serendipity, collegiality and brief commutes. In short, while the upper echelons may linger downtown much of the staff will choose, or will be coerced, into relocation.

The success of most, if not all, recommendations to reverse the tide in central cities hinge on safety: people won't come and won't stay if there is a pervasive fear of crime. Good, restorable buildings remain vacant or half-filled because they are located in a danger zone. Like other proposals to breathe new life into the central cities they call for cities to regain a suburban level of personal safety. Despite the sharp decline in crime rates in the 1990s, there is sufficient residual fear to render cities objectionable to middle-class people with small hostages to fortune, their children. In time, central city governments may adopt or condone the unthinkable: infringing on constitutional freedoms by keeping dubious people out of walled-off downtown gated districts with heavily policed internal security systems. In the late 1990s there were faint signs of urban revitalization in a number of cities with New York leading the way

The New York economy despite the "killer taxes and paltry public services" scathingly depicted in the city's conservative *City Journal* seems to be in a growth mode.[41]

> ...the city economy has benefited even from what might be casually perceived as the mother of all threats: the computer age. New York is home to "silicon alley," more than 1,000 small companies in the software business. And according to the calculations of Mr. Moss and his Taub Center colleague, Anthony Townsend, New York has the highest concentration of Internet domains of any large city. "Density still counts," Mr. Moss concludes:

> New York has also proved adept at recycling obsolete manufacturing and commercial space, making it profitable for some cost-squeezed businesses to stay in the city and for others to come in. The

publishing industry moved from midtown to lower Fifth Avenue. Big retailers colonized the lower part of the Avenue of the Americas. Wall Street moved north (while the advertising industry moved south) into TriBeCa. Led by the Walt Disney Company, Times Square is again a center for glitzy entertainment.[42]

New York City is only one example of the better news from the front. The steep population declines that reflected job losses and the exodus of the middle-class seem to have abated in a number of large cities including such stellar examples of urban decay as Cleveland, Newark and Detroit where the double digit population declines at the 1980-1990 decade fell to a more modest 1.5 per cent to 2.7 shrinkage between 1990 and 1996.[43] The exception to the rule is crime-ridden Washington, D.C. (10.5 per cent loss) and St. Louis and Norfolk, Virginia, while big population gainers are located in the West and Southwest (e.g., Phoenix, up 17.7 per cent between 1990 and 1996). Many troubled cities appear to have reached a kind of population equilibrium at levels far below their 1940-1950 peaks.

While Detroit is, at the moment, a worst case example, it is only one of many. In Baltimore, the vacancy rate in the central business district office space was 23 per cent in 1995. One observer suggests that the emergence of the new technology, teleworking, "could add years to the recovery" the time when growing demand finally catches up with vacancies.[44] Even this estimates seems highly optimistic.

Adaptive Reuse for Abandoned Downtown Office Buildings?

Just how likely is such a recovery in downtown office space, with new tenants and new owners bailing out the distraught building owners, mortgage holders and city government as similar upturns did in past years? In most cities the prognosis for older buildings is poor: many office buildings are obsolescent dinosaurs, in need of thorough and probably uneconomic modernization program. The alternatives are razing old buildings and attempts at adaptive reuse--probably as apartments. Assuming destruction is a last resort, what are the prospects for reuse? In Toronto, office buildings built three or four decades ago, are being converted.[45] Toronto, which has been described as New York City in the fifties is not a prototype: Toronto is a well governed, safe city.

In Detroit, the David Broderick's owners' hope is to convert it to an apartment house. But even with tax breaks, many developers are skeptical, skyscrapers rise in packs, a factor of urban land economics. How many people want to live on shadowy, narrow streets with no schools, parks or markets?

You can mothball buildings, but you still must pay taxes and renovate your old building when the market improves. By then, the building is even older and more outdated.

The most optimistic scenario calls for a gradual cycle of decline, in which cities lower tax assessments and landlords lower rents, opening towers to small business, government agencies and nonprofit institutions.

The least optimistic one involves wrecking balls and dynamite and an orgy of high-rise demolition.[46]

This phenomenon of office vacancies, cannibalizing to compete for fewer and fewer tenants and threats of abandonment is by no means confined to big cities. Stroll past your local strip development and count the empty second (and third) floor offices. Once there were doctors and lawyers, now if they're lucky, there are wig stores, tattoos, body piercing and judo parlors.

A generation past, in the precomputer age, Buckminister Fuller remarked that the downtown office area was a giant dormitory for typewriters. There was no life there after 5 or 6 p.m. Conversion of vacant office space for residential use can change the equation. As was noted, Toronto is well-along the road to adaptation of office buildings and by the mid-1990s it appeared considerable conversions were taking place in New York City's Wall Street where "hundreds of new apartments are planned for or are under construction in office buildings at the heart of the financial district."[47]

Toronto and New York City are not alone. Office buildings in Boston and Chicago are also in a conversion mode. What this suggests is that in large metropolitan areas there is a niche market for downtown office residences, numbering perhaps in the low thousands to possibly ten thousand. This is a small market but it can have a major impact on small neighborhoods. We can also anticipate that the new residents will not require public schools and will possess sufficient resources to ensure that such areas will be well policed with public police supplemented by private guards. In this context even minor successes receive considerable publicity. For example, a news story reported that since 1993 Savannah, Georgia leveraged private investment to revitalize Broughton Street and in the process assisted property owners in developing five new apartments in previously unused floor space.[48]

The question is: Can this kind of conversion be done on a massive scale--for instance,in downtown areas such as Detroit or Baltimore? Is the handwriting on the wall for the many cities with huge numbers of vacancies or see-through buildings, uncompleted for lack of tenants? The odds are formidable partly because of job losses and vacancies to come. While telecommunications technology is not yet the driving factor in troubled downtowns it may well become a major player in the coming decade. As one call-to-action puts it.

> Technological developments in information access and delivery have made business increasingly independent of location and have made it less necessary for many for many employees to travel to a central location to do their work. Census data show that about 15 per cent of Americans work at home full-time, with countless more doing an increasing percentage of their work from their residences. A twenty-first century in which telecommuting is the norm--in which many huge businesses maintain small central headquarters and vast networks of employees tied in from their homes by phone lines and computers--is no longer the stuff of science fiction: it is an onrushing reality. This trend poses an especially difficult challenge for New York City, which derives so much of its tax revenues from large office buildings and the commuters who work in them.[49]

If the vacant structure is still sound, and it is offered virtually free to developers some adaptive reuse may be financially feasible. Some buildings may fall into city ownership after non-payment of taxes. Others may be owned as non-performing loans, deadweight for cities, banks and mortgage companies.

If the acquisition price is low enough, it may be possible to see in some cities reuse for housing--rental apartments, subsidized affordable housing, single room occupancies, home-related offices, work studios, hotel and educational facilities. They could also be locations for suburban-style convenience and retail stores as well as traditional downtown shopping, restaurant, entertainment and cultural uses. Even performance-zoned; i.e., clean, noise-free, light manufacturing plants, might be potential tenants.

Much depends on the state of the city economy and the rebirth of investor optimism. For example, bad experiences with hotel overbuilding in New York City in the 1980s have been replaced by a rush to build and to convert-- in the late 1990s. In New York City:

> Hotels are often jammed to capacity these days and the average room rate has soared 30 percent to $174.71 since 1994, so investors and hoteliers are spending more than $1 billion to add nearly 6,000 rooms to the city's inventory of roughly 60,000.
>
> Nine of the hotel projects announced this year involve conversions of office buildings or warehouses, a process that is far less expensive than new construction...[50]

In short, given a soaring economy many well constructed, older buildings are candidates for adaptive reuse. The question is whether this kind of effective demand is or will be present in most central cities.

The Soft Path: Tourists, Convention Centers, Stadiums

A southern economic development specialist once said that its easier to pick tourists than to pick cotton. A number of troubled cities have taken the adage to heart, with mixed results. On a cost-benefit basis, the least defensible approach is the stadium. Open for only a limited number of days per year, stadiums require enormous subsidies to build and to operate. Oriole Park at Baltimore's Camden Yards offers a good case study of the phenomenon.

> Explicit or implied in forecasts of benefits to be reaped from hosting sports teams is the notion of a multiplier, i.e., an estimated figure which represents how much additional spending related to the team will spill over and encourage further local spending. The theory of the "multiplier effect" of government spending proposes that much of the initial expenditure will be respent locally by the recipients, which in turn will generate more local spending. ...Initial public spending will increase jobs, incomes and tax receipts, so that even if the project does not cover its own costs, the indirect benefits are expected to more than equal total outlays.[51]

In the case of Baltimore whose stadium is subsidized by the state lottery, it is clear that only about 1,500 full time equivalent jobs have been generated by the stadium. There is an equity issue involved in the financing: most of the lottery revenues come from two largely black and low income areas, Baltimore City and Prince George's County. As one state senator phrased it, the stadium "is being financed on the backs of the poor."[52]

It seems clear that as in the case of other sports investments the deciding element in the calculation is civic pride and prestige rather than prudent calculations of economic gains and costs.

The competition for sports teams has escalated to the point where small cities are engaging in bidding wars for minor league baseball. As William Fulton writes:

> ...like their metropolitan counterparts, minor-league towns are falling all over themselves to accommodate the owners, claiming...vast benefits for their communities.... In the end, what the minor league towns are paying for is nothing more than a designer name...In the case of minor league baseball, the price for tapping into that kind of identity is getting too high.[53]

The pioneering success of James Rouse's Faneuil Hall Market Place in Boston and Harbor Place in Baltimore led to a trend in festival market places, efforts that subsequently failed in Toledo and Norfolk. Convention centers have all too often attracted too few events to become self-financing. For example, Philadelphia's 1.3 million square foot convention center, hailed as the engine of the city's economic renaissance has fallen dismally short of its backers' promises...It was completed in 1994, three years behind schedule; by then its cost had soared to $522 million, double the original estimate. Worse, the predicted economic payoff--10,000 full-time and part-time new jobs and a steady flow of profits has not materialized. So far the center and hotel complex has generated only 1,100 full-time jobs.

Philadelphia's city officials dispute the figures and make the alternative clear: they have no choice but to seize on tourism because "there is not much else."[54]

A conservative assessment of the soft path to economic development might be mixed. New York and Miami, Los Angeles, Boston and San Francisco and other cities that are already attracting tourists and convention -goers might be well advised to move more strongly in this direction. Cities like Flint, mercilessly portrayed in "Roger and Me" might be advised to give tourism a pass. And there are a few mutants like New Orleans, badly governed, subject to high crime rates but still a great town for tourists.[55]

The Arts: Unlikely Soft Paths

Local support for museums symphonies, theater and ballet would seem to be an unlikely candidate for economic development. Yet, in the face of reduced federal funding, there are fifty-five to sixty officially designated cultural districts in cities across the country and local government financial support for the arts has grown by 4.7 per cent annually since 1991.[56]
We are reminded that:

> Cities have long been known for following the modish urban redevelopment concept of the day, only to find their tomorrows beset by the flaws in those ideas. Urban renewal left vast empty tracts. New highways wound up speeding the flow out of cities rather than traffic into them. Downtown office development bred design sterility and nighttime ghost towns. Whether reliance on the arts will have a similar legacy in many locales is an open question.
>
> Many fear that without solid indigenous programming, an interested audience to build on, and links between the projects and the fabric of the city, the grandest of the arts projects can turn into white elephants. Indeed, since Lincoln Center opened in New York City in 1962, more than one city intent on replicating its success has built an arts complex that remains underused and cut off from the city.[57]

So far the returns seem mixed from a financial standpoint but optimism endures. In conjunction with and in contrast to the exhibition halls, athletic stadiums and convention centers that cities have sponsored, the new arts focus seems designed to appeal to the niche demographic markets that cities attract--young adults, empty nesters and some immigrant populations. In addition, there is always the hope that this approach will stimulate tourist expenditures, raise local morale and possibly, regenerate blighted neighborhoods.

Gambling

In recent years, the proliferation of casinos and resorts has led to demands--promises--that economically troubled areas could use the

revenues from gambling to finance schools, create jobs and in general, enjoy prosperity without pain. The essential premises behind this reasoning are that
- people gamble and that if your city doesn't get its share, others, less fastidious places will certainly do so;
- gambling money will be fresh, derived from tourists and out-of-towners, not from poor residents spending their food and rent money at the casinos and;
- the modern "gaming" industry is free of the taint of organized crime and operates like any other business.

Given America's fascination with wagering, it is not surprising to find casinos cropping up throughout the nation, in cities, on riverboats, and Indian Reservations. In some ways, it is surprising that there aren't more of them, particularly in hard-pressed industrial cities. What is becomingly increasingly clear is that this is not a win-win avenue to prosperity. Gambling addiction is widespread, and social disruption is an inevitable byproduct of seductive slots, cards, wheels and tables. Like Latin and Caribbean nations involved in the drug trade, there is almost always, the discovery that the vice is not just an export--it also caters to the home market.

On occasion, what is almost as bad is heavy municipal investment in a casino that flops. New Orleans is a case history.

> It looked like a sure bet: The world's largest casino, a Las Vegas under one roof in the heart of New Orleans, a city with 10 million tourists a year...But today, the Casino, larger than a Manhattan block, sits empty and unfinished...Last November the $823 million Harrah's Jazz Casino filed for bankruptcy protection. Thousands of wage earners were abruptly laid off.[58]

New Orlean's experience has caused considerable reflection in Miami, Detroit, Philadelphia, Chicago and New York City, all of which were studying the possibility of large scale casino gambling.

An authority on the gambling industry summarized, the results of Harrah's aborted effort: If you're picking local people's pockets you can't call it economic development.[59]

Sixty percent of the casino's customers were local. In short: casinos squeeze the poor. There is no guarantee they will prosper when competition for the gambler's money is fierce. And when they don't work, nobody wins: no jobs, no tax revenue, blighted investment.

Movies

The problem with government efforts to promote non-traditional programs for economic development is that they tend to be effective only when there is already a substantial private sector base.

Moving picture production is a good illustration. While there are outcroppings in the South and Midwest, the pools of talent, financing and infrastructure remain firmly bicoastal--New York City and Los Angeles with small subcenters elsewhere.

This is not a small contribution to the economy. Production, animation and others aspects of the industry helped take up some of the slack when aerospace dehydrated in southern California. New York City's Commission of Film, Theater and Broadcasting estimates that $2 billion was spent directly in the city on film and television work and that 78,000 city residents are professionals in the industry.[60]

(These generous estimates represent an attempt to counter residents protesting temporary, disruptive use of city streets, parks and sidewalks.)

City Farming

One often overlooked remedial program is city farming. At first glance this seems like an oxymoron but the reality is far different. Research shows that more farm-related jobs are found in Madison, Wisconsin than are offered by the state university. Focus on intensive agriculture--vegetables, orchards, vineyards and herbs--in fields or under glass or plastic can generate substantial amounts of produce up to fifty jobs per acre. There are now over 2,000 farmers' markets that rely on high-quality local (and regionally) grown agricultural output.

This path offers substantial supplemental income for city families, an alternative to illegal activities and the prospect of neighborly cooperation. In Detroit, over one-third of the land within the city limits is vacant. Similar sizable acreages are available in other cities, especially those with heavy population losses. (Some of this land may require brownfield treatment.)[61]

An End to Gentrification?

In the late 1960s, dozens of cities were celebrating an apparent reversal of the middle- class flight from cities. Boston's South End, Baltimore's Fell's Point and Washington's Capitol Hill were the targets for middle- class settlement, some from originating suburbs but mostly generated in the central cities themselves. They captured a small proportion of middle- class housing demand in metropolitan areas and the result was a combination of rising real estate prices and cries of pain from some of the poor, particularly blacks, who saw the influx of the well-off as part of malevolent master plan to reclaim the cities from the poor and to banish low income renters to close-in suburbs. For a time, gentrification seemed so unstoppable that an occasional optimistic planner could call for distributing the benefits generated by settlers evenly; i.e., better public services, instead of allowing the goodies to concentrate in some neighborhoods.

In recognition of the fact that the issues confronting central cities are not engineering problems but people problems, some cities have deliberately or tacitly encouraged the influx of new people--better behaved, sober, tax-paying, hard-working--as a counter balance to their dysfunctional residents. Gentrification as it is called, represents a recapture (or retention) of middle-income residents by cities.

For the most part this was a spontaneous movement with the cities playing the role of bystanders or moderate helpers. A few cities like Baltimore stimulated the process by selling hundreds of city-owned abandoned units to new owners for $1 as part of a homesteading program. The apparent success of gentrification in transforming rundown housing prompted calls in a number of cities to restrain gentrification on grounds that it was pushing the poor out of neighborhoods. Renters in particular were hard pressed to find comparable housing. In contrast, homeowners benefited to varying degrees when newcomers bid up the price of their properties.

Come the mid-1990's and the picture is far different, much less rosy. To a degree, gentrification was always more hype than substance. Baltimore's well-known homestead program, for example, redeveloped less than 1,000 units out of the city's total of 225,000. Perhaps if the trend had continued for another decade or two, central cities would have become more evenly balanced between the affluent and the poor, but it was not to be. The fact is that while real estate values have leveled off in established upper- income city neighborhoods by the late 1990s, many of the reclaimed gentrification areas have run into a brick wall. Located in or near high crime areas, their residents frequently find it difficult to find a buyer after the latest mugging, break-in or murder. Finding a buyer, even the legendary "Greater Fool" that sometimes rescues the lesser idiot from what turned out to be a reckless purchase is not feasible.

Salvation Through Immigration?

A major source of city regeneration is a return to earlier tradition. In the nineteenth century and early part of the twentieth century, many cities were buoyed by wave after wave of immigrants. While there was considerable friction and fears that America was being mongrelized and its jobs and businesses were being lost to aliens the net results were mostly benign.

Despite exposure to years of hardship and the occasional danger of robbery or worse by slum predators, immigrants to central city neighborhoods keep arriving for much the same reasons as the newcomers who came a century or two ago.

> Not only were American wage rates high, but you kept what you received...no conscription. No political police. No censorship. No legalized class distinctions... No one called anyone "master." Europeans learned these things from relations or neighbors who had already taken the plunge--letters which circulated round the village or were read aloud to rapt gatherings around the fireside.[62]

One special characteristic of many immigrants, past and present, is the high priority they give to home ownership. A HUD report asserts that "the immigrant surge in the 1980s and 1990s will be a substantial source of future home ownership demand.[63] This is by no means a new phenomenon. At the turn of the century German, Scandinavian and Italian immigrant levels of home ownership were substantially higher than ownership rates among native Americans.[64] In contrast to the problematic Yuppies market, a substantial amount of vitality and urban renaissance can be attributed to foreign immigrants. Loosening of immigration restrictions in the past several decades has resulted in a truly massive multi-million population influx that has concentrated in six gateway cities: Los Angeles, New York, Miami, San Francisco and Chicago.[65]

Owners of central city property might look around them before they bail out at a rock bottom price. Is the building in the path of a wave of new immigrants? Are foreign language signs sprouting on storefronts, a block or two away? What are the trends in comparables for property sales in nearby immigrant enclaves? Are immigrants spilling over into your neighborhood?

There have been occasional outbursts in the 1990s against immigrants in the time-honored fashion of blaming newcomers who are allegedly responsible for urban problems; from worsening slums to high rates of native unemployment, from overburdened schools to punishing tax burdens. The data suggest otherwise. High immigrant cities have lower unemployment rates and higher incomes than cities barely touched by significant numbers of newcomers.

New York City offers a prime example of salvation; in part, through a surge in immigrants.

> While New York's attraction to immigrants is a mixed blessing--the city's schools are crowded with their children--immigration's constant renewal of the city's energy has almost certainly outweighed the costs.
>
> For one thing, immigrants have filled the void in many neighborhoods left by deindustrialization and corporate downsizing. The city's foreign-born population, just 18 percent at its nadir in 1970, reached 28 percent in 1990 and is still rising rapidly, with Dominicans, Russians, Chinese, Jamaicans, Guyanans and Indians leading the charge.
>
> For another, immigrants reinforce the city's strengths as an international financial center. The presence of large, distinct immigrant communities makes it easier to house and school foreign nationals, as well as to find local labor that speaks foreign languages. And "unlike Los Angeles, New York has strong ties to Europe as well as Asia and Latin America," Mr. Hormats of Goldman, Sachs said.
>
> Immigrants also create natural markets for ethnic products that give manufacturers a chance to expand to the rest of the country--and the world.[66]

An official program to replace fractious natives by law abiding immigrants is a political impossibility but what we have in practice is an unofficial policy that accomplishes the same result. In different ways communities have offered a warm welcome to constructive newcomers, helping along the process of invasion and succession that is a feature of urban living.

Real, honest-to-god immigrants from overseas was the group that Peter Hall hoped to attract to his proposed city enterprise zones that could confer accelerated progress toward citizenship on overseas migrants who pony up substantial investments to generate jobs in slum neighborhoods. This group has been selectively culled by Canada, Australia and elsewhere, places that have been creaming nervous investors from Hong Kong for over a decade.

The Economist reminds us that immigrants are here for work; not welfare.

> Many Americans fret that immigrants endanger their jobs. They should thank them instead. Latinos, in particularly, have played a big role in keeping America's economy humming. Almost 40% of those hired for new jobs in 1996 were Latino, even though this group makes up only about 10% of America's population.[67]

The typical immigrant is not an investor but is much more like the turn-of-the-century steerage passengers; poor in money, often ill-educated but rich in discipline, energy, family cohesiveness and resistant to the temptations of alcohol, drugs, crime and indolence. It is this kind of traditional immigrant in the tens and hundreds of thousands that has transformed, for the better, entire city neighborhoods in New York, San Francisco, Miami, Washington and other cities. Encouragement for industrious newcomers in the form of gentle forbearance from the federal Immigration and Naturalization Service, the availability of low-cost housing, free English-as-a-Second language courses, cheap vocational education training and effective policing can make a big difference.[68]

> Almost all in time find work--often jobs no other people (even low-skilled black Americans) want. Many join existing firms, start their own businesses, however modest. As a result parts of New York are seeing the emergence of a new immigrant "middle class that, at least in terms of income, matches indigenous Americans living in the

same neighborhood--and pays taxes to match. Newsstands and bodegas may seem humble, but they are still the stuff from which immigrant dreams are made.[69]

What can poor immigrants contribute? We have the example of the Vietnamese manicurists, 16,000 of them in California who have driven the cost of manicures down from $60 to $9 in a few years. Weekly manicures are now affordable. Why manicurists? After two months of training, a little English and $2000 to $3000 in equipment is all that it takes. In contrast, Cambodian-Americans have found a niche in doughnut shops, Indo-Americans in motels and Chinese-Americans in gas stations. Pakistani immigrants have made successful inroads into New York City's discount stories.

Buoyed by a surge of immigration into the city in the 1990s, newcomers are going into business at a furious pace, creating new employment niches for themselves by bringing friends and relatives with them...

'Usually two or three brothers or cousins live together, work separately and save their money to open a store together...' The initial investment required to open a store: $50,000 to $60,000.[70]

What do these businesses have in common? Like the groceries owned by South Koreans in Washington, these occupations demand family support, long hours and hard work.[71] Natives don't seem to be up to it. Like their earlier forebears, today's immigrants have the necessary stamina, mutual trust and frugal discipline.

Some of the most dramatic examples of cultural changes in a relatively short time have been among the immigrants to the United States from Eastern and Southern Europe and their descendants. Although, most were uneducated and barely literate during the era of mass immigration--and indeed often resistant to education for their children--Southern and Eastern Europeans eventually became as educated as other Americans and as well-represented in occupations requiring education, such as professional, technical and managerial positions.[72]

The kind of rescue-through-population-change suggested in the text is a living reality in Lefrak City. Consisting of twenty, 18-story buildings on a forty acre site, Lefrak city contains 25,000 people in New York City's Borough of Queens. In the mid-1970s, a federal housing discrimination suit accused the Lefrak organization of discriminating against blacks who were kept on a waiting list while white tenants were moved to the head of the queue. The rationale: Lefrak city was 8 per cent black, 82 per cent white, and the management wanted to preserve the racial balance. After the organization yielded to federal pressure, the black percentage increased to 67 per cent by 1980 and 79 per cent in 1990. (This might be classed under the heading of unanticipated consequences of good intentions gone awry.)

By 1980 Lefrak City was largely poor with rising crime, broken benches, damaged locks on entrances and graffiti on corridors and in stairwells.... But things began turning around in 1991. Management began evicting more problem tenants.[73]

Aside from the removal of trouble-causing tenants, Lefrak city began to attract two groups of family-oriented, religious tenants, one stream from West Africa, mostly Moslems and another Jews from central Asia. Signs of the new age: a new mosque and a reactivated synagogue. All told, an estimated 5,000 tenants came from these two groups with new arrivals being added each mouth. The guns and drug dealing that left the project "teetering on the edge of collapse" are gone. Lefrak City has stabilized.

A footnote to city upgrading-through-immigration is the unheralded beneficial impact of Caribbean migrants. The West Indies have been the source of a significant number of education-oriented, entrepreneurial, home owning, achieving immigrants. Between 1990 and 1994 over 200,000 of New York City's 563,000 new immigrants came from the Caribbean.[74] Thomas Sowell has identified the distinct advantages enjoyed by and the special characteristics of West Indians.[75]

> In the public schools the mass of immigrants has been critical in raising academic standards. Earlier this year, when the Board of Education released reading and math scores for public high schools, three of the top Brooklyn schools "belonged" to immigrants: Midwood, Clara Barton and East New York Transit Tech.
>
> Midwood draws immigrants from several Russian enclaves nearby, but also has Brooklyn's only bilingual advanced science program in Haitian Creole. The success of Clara Barton and Transit Tech is even more striking. Both are considered vocational schools, yet send more than half their graduates to college. Both are dominated by West Indian immigrants.

'Immigrants aren't the sole reason for the school's turnaround, Principal Lazarus says, 'but they're an important factor: the tremendous work ethic, the way they are so focused on achieving a goal. And the family structure is very strong, so there is a high regard for education as a vehicle to advance themselves.'

In black Brooklyn, reform followed immigration, not the other way around, and is playing an integral role in restoring the public school system.[76]

The data suggest that neighborhoods that have received substantial numbers of most immigrant groups have benefited. Between 1970 and 1995 almost 600,000 people, almost a tenth of the city's population, left for the suburbs or other destinations. Their places were taken by almost 500,000 immigrants. The top five groups seeking citizenship: Dominicans, Chinese, Jamaican, Guayanese and Haitians. Note that four out of the five originated in the Caribbean.

With a population fluctuating between seven and eight million for the five decades since the 1940's, New York has not suffered the huge percentage losses that have afflicted cities like Detroit and St. Louis among many others.[77] But the population mix is a lot different than it used to be. For one thing not all the outmigration is confined to the well-off. Puerto Ricans return to the island, forming a new composite," New Yoricans." Black southern-born persons retire South or send the children to low crime areas to live with grandparents or other relatives.

By one estimate, in the mid-1990's half the population of New York is foreign-born or the children of foreign born --a pattern that was common two generations back but a rarity in the 1990s. In New York, affluent whites are still departing but

> Puerto Ricans and native born black citizens--who constituted the greatest wave of newcomers in the postwar decades--are now leaving the city in proportionately larger numbers than whites.... The Planning Department estimates that about 40 per cent of the city's black residents are foreign-born or descended from a foreign-born parent... Except for Dominicans and Mexicans, all the foreign-born groups in the city have a lower percentage of people in poverty than the city's average of 19 per cent.... The median income of foreign-born black residents is $28,000 compared with $22,000 for native-born black residents.[78]

Are we likely to see more cultural changes in New York and other cities that are experiencing a similar influx of immigrants---San Francisco, Miami, Los Angeles in particular? To a degree, this has already happened with a cultural transformation that as noted, has rehabilitated entire neighborhoods in these immigrant gateway cities. In the long run we are in a race between the familiar challenge of mainstream assimilation into the working-class and middle- class of young people, the children or grandchildren of immigrants or their drifting into the dysfunctional underclass, preying on their neighbors and the rest of the city. There is the long term tradition of generations of immigrants using the city slums as a transitional staging area on the way to the suburbs, a pattern that is in full bloom in many metropolitan areas. The question is whether there will be a sizable residual, stable population in lively immigrant neighborhoods that make cities once again attractive to industry, commerce and the middle- class.

One major and almost immediate contribution of immigrants has been their positive influence on housing markets. Like their predecessors around the turn of the century, the new immigrants display a marked predilection to save for home purchases. As noted in *The Wall Street Journal*, this source of home buyers counterbalances the decline in baby boomer demand. Research indicates that Asian immigrants are already in the process of buying up new single family housing while Hispanics

> ...have emerged as buyers of condominiums, townhouses and rent small apartments, striving to buy something bigger...typically Hispanics arrive alone and gradually bring along extended families.[79]

Palestinian immigrants run many mom-and-pop grocery stores in San Francisco, Detroit and elsewhere. *Koreans* successful in operating small markets in big cities are opening bigger stores and dispersing to the suburbs. *Vietnamese* operate small parlors for trimming and polishing nails. (And are moving into hair cutting and hair care.) Barbadians specialize in short haul trucking especially in New York City. *Chaldeans* (Iraqi Christians) own many convenience liquor stores in the Detroit and San Diego areas. *Chinese* aside from restaurants, are the key players in the toy wholesale business in Los Angeles. *Cambodians and Vietnamese* operate doughnut shops, particularly in southern California. *Indians* own 46 per cent of American's economy hotels. they are now buying major hotels, Sheraton, Radissons and Hiltons.[80]

Like their predecessors for the past two centuries, most poor immigrants head for the cities. One critically important group, seems to have a suburban bias. This is the highly trained professional category, engineers, information specialists, many from Asia, others from Europe who gravitate to Silicon Valley and other high technology development clusters. India sends between 12,000 and 15,000 of its annual crop of 50,000 information-technology graduates to the United States.[81]

A few of these well paid immigrant professionals may end up in New York City's "Silicon Alley," or similar concentrations in Austin, San Diego or Portland, cities which manage to buck the trend by remaining attractive to upscale people. For most cities, any sizable influx of immigrants is unlikely and attracting numbers of sophisticated investors or professionals is simply not in the cards. They may staff their hospital emergency rooms with foreign doctors and nurses and their local colleges with foreign mathematics teachers but that's about as much as we can expect.

Happy Indians, Restless Natives

In the summer of 1997, *City Journal* published two articles that underscored the challenge and change taking place in many cities: "Bombay on the Hudson" depicted the mostly successful integration of 500,000 immigrants from India, Bangladesh and Pakistan into a city of less than 8 million. "Vibrant Indian neighborhoods" enliven every corner of the city.[82]

The same issue of the publication laments "Gotham's Workforce Woes." These turn out to be very much in line with William Julius Wilson's depiction: lack of basic skills in numeracy and literacy, speech patterns well outside the mainstream; a lackadaisical approach to work, hours and business norms and a profound resentment of authority. For this reason, there is a dearth of good entry-level workers.[83]

The result is large scale unemployment and seething bitterness on the part of the native born--with a large black component--side-by-side with a rapid rise in income and status on the part of new immigrants. The torching of South Korean stores in the 1992 Los Angeles riots is only one manifestation of this resentment.

The partial replacement of the dysfunctional native born by immigrants is not limited to the more exotic strains of newcomers. In the Los Angeles--Beach area, Latinos now buy more than half the homes in Los Angeles County, own a quarter of the businesses and are making run-down neighborhoods vibrant, according to Henry Cisneros, former Secretary of Housing and Urban Development.[84]

Given this important role in municipal rescue operation there is two critical questions. The first is whether the left-behind natives will accelerate their progress into the mainstream or will form a troublesome population of permanent outsiders. The second is whether the new immigrants will celebrate their success by moving to the suburbs as they rise in incomes or whether they may remain in the city neighborhoods they helped to revitalize.

With the crime problem less vexations the key is public school quality. since tuition in private schools run from $5,000 to $15,000 a year per child. Unless here is a substantial improvement in the big city public school systems. In the next decade, there is every likelihood that the new immigrants will join their predecessors in the middle- class and relocate to suburbs. Indeed, there is evidence that this is already taking place and that, moreover, the suburban immigrant colonies are being augmented by a stream of new arrivals direct from the old country.

OTA (Now Defunct) and Central Cities

Official government publications find it difficult to play the role of undertaker for afflicted populations and areas. Triage cannot be adopted as official policy. The cool reception given to the Carter Administration's 1980 urban policy report that implied a generational write off for northeastern "rust belt cities" is one example. The controversy surrounding the Moynihan 1964 report on the troubled Negro family is another[85]

Even an implication that cities are being written off is unacceptable. For example, one of the difficulties in designating growth centers in the Economic Development Districts of the 1960's was that the non-designated cities were, by omission stigmatized, marked as hopeless cases.[86] The usual politically approved compromise was to designate all of the contenders for growth center laurels rather than singling out one or two champions with genuine potential.

Much the same experience accrued in the 1960s and early 1970s with the Model Cities program when the number of demonstration cities increased from six to sixty to 150. Political realities dictate that every area must look like a winner.

Is this an adult version of sandbox politics? A large kindergarten class in New York mounted a revised version of a traditional fairy tale for admiring parents: "Snow White and the Twenty-seven Dwarfs." Every tot was on stage, none was a spectator This was a valuable lesson for the children as an early introduction to the real world of log rolling, influence swapping and pressure groups.

It is not surprising that the 1995 Office of Technology Assessment study on the future of metropolitan America found much prospective good news along with the bad for central cities. True, OTA recognized problems for the urban poor created by a combination of geographical mismatch (new jobs in the suburbs) and a skills mismatch (new jobs require more skills.) They saw, accurately, new strengths from "vibrant immigrant enclaves" and from pockets of gentrification. OTA points out that downtowns still have prestigious addresses, that good public transit systems provide access for large pods of lower-level employees to the CBD. Moreover, OTA believed feels the glut of office space in central cities may have a silver lining; rentals at fire sale levels may come close to rent levels in the suburbs. Thus, there may be a future for low-wage industries like apparel/manufacturing in Los Angeles and New York.

But the report underscored the disadvantages: high tax rates, rigid zoning regulations and a poor image compared to suburbia caused by "decay and crime." OTA recommended remedies? Business improvement districts which levy special taxes on commercial property for extra security and sanitation to curb "crime and grime" enclavism on the Chicago model where local officials have considered a proposal to close off a number of streets in an existing industrial area to create an industrial park with one secure entrance.

OTA also offers cheering data on the strength of global cities; e.g., New York, San Francisco, Los Angeles, and Chicago, where face-to-face contacts by lawyers and bankers add significant strength to special assets--culture, entertainment, sports, design, medical facilities, higher education. But there remain fundamental defects pointing the way to further trouble for the strong and to very serious problems for the weak:

Crime rates are likely to reach "intolerable levels." (In this respect OTA missed one bit of genuine good tidings, the decline in city crime rate in the 1990s.) Too many public school products are woeful unprepared for the world of work, there are high taxes for poor quality services and there is an unlikelihood that there will be effective state or national policies for reconfiguring public programs to slow suburban growth in favor of central city locations. Moreover any such efforts that may be in progress or in prospect are not apt to change the fundamentals--immigrants aside, suburbs are more attractive places for families. Chronic misgovernment is one important element in the suburban exodus, a failure discussed mostly by indirection in the OTA study.

Under the heading of paying more and getting less for municipal services, we have the stellar case of Washington, D.C. compared to its suburban neighbor.

> ...it takes an average of 20 minutes--twice the national norm--for an ambulance with paramedics and advanced life-support technology to arrive, according to an audit of the D..C. Fire and Emergency Medical Services Department released last week. And when paramedics arrive, they have none of the critical medicines-- morphine, Valium and Adenosine--routinely carried by their counterparts in the suburbs.
>
> D.C. firetrucks and ambulances with basic life-support equipment and medical technicians take an average of five to ten minutes to respond, also longer than the national average...
>
> "It seems unjust that someone suffering a heart attack on the east side of Western Avenue [in Washington] should receive care that is sub-optimal by design, while someone suffering a heart attack on the west side of Western Avenue [in Maryland] would be afforded better care," the audit concluded.[87]

The deficiencies in emergency medical services reflect a broader pattern of municipal mismanagement. For example, the sad condition of Washington's public school system is reflected in the estimate by one school administrator that in some upper- income Washington neighborhoods 80 per cent of school age children attend private schools.[88] The pattern of poor management seems to encompass every city agency. It was found that in 1997, "There is only one working photocopier in the entire Transportation System Administration, an agency with an annual budget of 19.6 million and 444 employees."[89]

The *City Journal* quotes a former member of the mayor's cabinet on the root cause of the mismanagement problem in Washington and by inference to other cities; government wasn't seen as providing a service but providing an income.[90]

Redesigning Cities

For over a century most of the efforts to improve, adapt and save cities have focused on architectural and engineering remedies. There are several reasons for this approach. The simplest one is that building structures is doable in contrast to embarking on risky and often, unmeasurable social programs. Moreover buildings are tangible, politically saleable commodities. In contrast, people programs involve intervention among what Victorians called "the undeserving poor," persons burdened with the stigma of social and political leprosy.

Alexander Garvin's thoughtful book subtitled *What Works and What Doesn't in American Cities* offers a long list of successful ventures in constructing effective park systems, cultural centers, shopping centers, pedestrian malls and housing construction and rehabilitation. There is a chapter on neighborhood development and another on effective historic preservation.

Garvin gives special attention to amenity-rich cities like Portland, Oregon and Minneapolis and gives them deserved congratulations on a job well-done. What is missing is a searching inquiry into the people problem. Indeed, the index contains no references to critical problems that make cities unpleasant or intolerable: nothing under crime, corruption, race riots, deteriorating schools.[91]

This apparent oversight is by design. Garvin indicates that good project planning is not a remedy for social problems, and his clear focus on successful marketability incorporates by implication the impact on considerable serious urban dysfunction.[92] In general, he seems to feel that good design, proper location, effective political leadership and urban entrepreneurship can provide the answers we need. In the case of a city like Portland which appears to some like a middle American theme park afflicted by only the most minor blemishes, this is clearly the case, but in the problem-wracked cities good design, even in the form of protected enclaves offers small consolation. It is reminiscent of

Luigi Barzini's verdict on the subsidized industrial developments in southern Italy that failed to take root and generate new growth--"Cathedrals in the Desert."[93]

One approach to redesign is razing and replacing troubled high rise-buildings containing concentrations of dysfunctional tenants. For almost a generation, high rise public housing has been the target of planners who charged that such structures were unsuitable for low income families, particularly AFDC welfare recipients who provided recruits for gangs and drugs. The demolition of the Pruitt-Igoe project in St. Louis in 1976 was a portent of things to come. High-rise buildings were demolished in Baltimore, Chicago, Newark and other cities in the 1980s and 1990s and a total of 100,000 units were slated for razing by the year 2000 at an estimated cost of $2.5 billion.

Baltimore is a leader of the pack. The city plans to raze forty buildings in four public housing projects making it the first city to eliminate all public high rises. (New York City is the exception with almost 3,000 apartment buildings with 600,000 residents, the city has no plans for demolition.)[94]

The destruction of high rise public housing and its partial replacement by townhouses is the latest remedial program based on an assumed correlation between bad design and bad behavior. Oscar Newman, among others, charged accurately, that high-rise structures all too easily become uncontrolled havens for delinquents and criminals and that at the very least they should be retrofitted to eliminate blind corridors and shrubbery concealing predators.[95]

Total demolition is based on the twin premise that high-crime high-rises would be replaced by low crime low rises. Such expectations may be unjustified. The film *Boyz in the Hood* portrays the crime-ridden south central area of Los Angeles where the single family housing could easily pass for a working-class neighborhood in any city.

This underscores a basic truism: engineering problems are easy. We know how to build good roads, schools and housing. People problems are hard. We don't have sure-fire methods of guaranteeing well-behaved neighbors, studious pupils, disciplined workers. There has been a good deal of confusion about linkage between slum buildings and the behavior of many slum people. Experience suggests if we change the population mix, for example by encouraging (or at least not discouraging) settlement by Koreans, Vietnamese, Barbadians or Cubans, neighborhoods change for the better.

Remaking the City: Finding the Niches

In an era when corporations are no longer embedded in communities, cities have discovered the broader applicability of Rosabeth Moss Kanter's admonition to workers: "Job security is no longer in the national business lexicon." Neither is any assurance that the corporation will linger as a contributor to the tax base, to civic leadership and to the local business scene.

Kanter accepts the culture of bottom-line business efficiency as a given. In response to 40,000 layoffs in the corporation she feels that global competitors like AT&T need the flexibility to pursue resources and markets anywhere they can. In the short run, this may seem uncompassionate, but a corporation that doesn't behave this way risks failure, jeopardizing many more jobs in the long run.[96]

Her solution: Let cities find a niche as thinkers, makers or traders. Examples of thinkers, Boston's Route 128, Silicon Valley, Austin, Texas and Northern Virginia. Perhaps Houston. Creative, innovative, agile, these are areas on the cutting edge of innovation.

In her classification cities that are "Makers" are high quality, manufacturing centers like Spartanburg and Greenville, South Carolina and reborn Cleveland. "Traders" include cities that are located at a crossroads of cultures like Miami.

How to strengthen a niche? Often a positive business climate in excellence in higher education, infrastructure, telecommunication, in public schools and amenities like parks. Needed: active business collaboration to fashion and implement a successful strategy.[97] What's missing from this picture? One problem is the breath and commitment of corporate leaders; most executives are deeply committed to personal career advancement rather than long term civic citizenship in a single community. Lawyers, doctors, utility executives and small-to-medium-sized business owners and executives may fill much of the gap but finding high quality leadership in an era of diminished loyalties and high mobility presents a serious challenge.

A second difficulty is the large number of troubled areas that can't or won't pull off a comeback. New York, and San Francisco and other cities have a lot going for them and are likely to make the grade but there are literally dozens of urban areas with little prospect of emerging from present miseries. One can have little confidence in cities that can't seem to put their house in order. City residents have reluctantly gone private to provide even basic services. In New York City's Charlotte Gardens, a small thriving single family enclave in the South Bronx, residents have reacted to the woes of the public schools by enrolling their children in parochial schools.[98]

The third problem is the widening gap between central city and suburbs, between winners and losers. A simple prescription: "modernize poor quality local public schools" faces staggering obstacles. Well-off suburbs are more than happy to attract high quality development and leave the troubled cities to sort out their own vexations.

In this context of past problems, long term trends and the likelihood of more troubles ahead, it is hard to agree with former HUD Secretary Henry Cisneros that the "free fall" of American cities is over and the difficult climb back has begun.[99]

Cisneros is among the optimists who see the cities stuck with the residue of the sweeping economic changes of the postwar decades now, "headed for a comeback." Toxic waste, empty lots, broken down factories and the black underclass are the legacy but a combination of welfare reform to attack the intransigent problem of the underclass and telecommunications and specialized manufacturing will transform the cities. In his view, "Detroit, down from 15 automobile plants to just three, shrunken from 2.2 million residents to just about a million, a virtual basket case..now seems poised for a revival."[100] However, Cisneros is careful to hedge his bets by stressing the self-help and civic leadership prerequisites for substantial improvement.

Roger Lewis, another more cautious, optimist does see faint signs of hope in Washington, D.C. But by implication Washington, like other troubled cities still repel middle-income residents and businesses by a combination of high taxes, an onerous regulatory environment, lack of public safety a poor public education system, dirty streets, decaying public buildings and often inoperable utilities.[101] If as Lewis suggests, major reforms, across-the-board in all these vital areas is a significant turnaround the outlook is grim. (One devastating statistic: In 1996 only 6 per cent of Philadelphia's high school students tested as competent in reading.[102]

Crime the Killer

If there is not much hope for placement of industrial sites in high crime areas, the same holds true for office buildings, commercial areas and housing rehabilitation. The question is why, if criminality has long been a feature of the cities, is it now lethal? If crime rates have been deplorable for well over a century, why did cities flourish in the 1890s through the 1950s? The answer seems to be metastasis. In earlier generations parts of every large city were lawless dens of iniquity, places where even the police went warily and on occasion, with black bags wide open for payoffs. But the popular culture of the twenties and thirties; i.e., the movies, picture a time when young couples could stroll at night in the parks without fear, where young children could be sent off on early evening shopping errands, cities of fear-free, stoop-sitters, saunterers and street musicians. Lewis Mumford offers a bitter contrast of past (i.e., 1910-1950) to the 1970s and early 1980s. In New York City:

> "Crime," he said, "used to be confined, like a carbuncle, to certain self-enclosed areas, like the Bowery or Hell's Kitchen. Such quarters had not yet poured their infection into the whole bloodstream of the city... For one thing, it was possible for men, women and children, even when alone, to walk over a great part of the city, and certainly to walk through Central Park or alongside Riverside Drive at any time of the day or evening without fear of being molested or assaulted."[103]

Peter Hall reminds us that it is not only a New York or Washington or Chicago problem alone but a well-near world-wide one. He sees a

> ···mountain of urban crime, and above all violent urban crime, that in the late twentieth century rose almost visibly, like some erupting volcano, to threaten the fabric of social life in every major city of the world. It was, indeed, a twentieth-century plague. And its causes were as mysterious to the afflicted, as those of the Black Death to the hapless citizenry in 14th century London.[104]

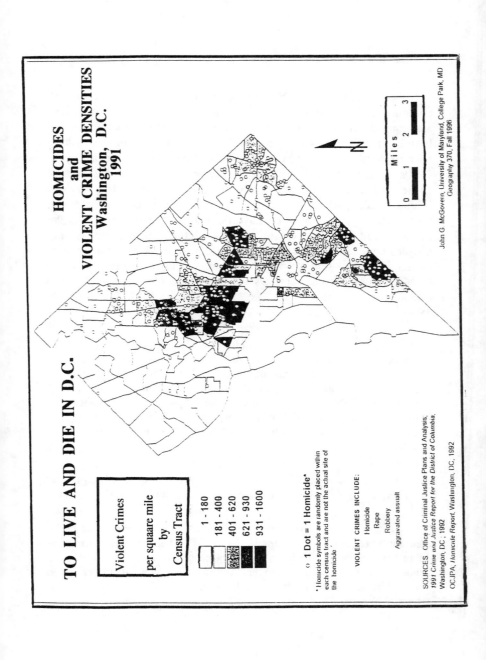

Why the Surge in Crime?

Observers usually point to a variety of reasons: a society that has "defined deviancy down." Imbibing liquor, pornography and homosexual activity used to be criminal offenses. No more. There is the everyone-is-a victim assumption. The basic charge by conservatives is that the criminal is no longer regarded as a moral leper, someone beyond the pale, stigmatized for his sins by an indignant society but instead is seen as a victim, the end product of a sick society. Defense attorneys enter pleas (sometimes successful) relating to previous child abuse, scars from racism, mental stress, brain shrinkage, excessive consumption of Twinkies. From this viewpoint, what we got in return for all the bleeding heart programs of the 1960's- the welfare programs, the light sentences, the coddling in prisons--is rampant law-breaking. In contrast, liberals have been on the run, seemingly incapable of mounting an adequate defense even of cost effective programs like the in-prison education that reduces recidivism.

In context, the weakening of neighborhood ties, the decline of active church memberships, the splitting of families which is so prominent a feature of city life in the 1990s has roots that go back to earlier eras. When the sociologists at the University of Chicago investigated city problems in the 1920s, Professor Robert Park concluded that

> Cities, and particularly the great cities, are in unstable equilibrium. The result is that the vast casual and mobile aggregations which constitute our urban population are in a state of perpetual agitation, swept by every new wind of doctrine, subject to constant alarms, and in consequence the community is in a chronic condition of crisis.[105]

Crisis or Not?

Chicago in the 1920s was like New York City when Lewis Mumford was cutting a swathe through Greenwich Village and finding the city a safe place for strollers: crime was highly localized. The question is why so much crime in so many parts of so many cities and why is it so pervasive in the 1990s? And here we are baffled. Why was the poor black family relatively stable, as for example, in its proportion of out-of-wedlock births from the Civil War through the 1930s only to reach near meltdown by the 1990s?. One in five District of Columbia blacks were born out of wedlock in 1881, one in five in 1939. By the mid-1990s, the proportion was four out of five. We know that kids who grow up fatherless are the population reservoir for the prisons and for the dire predictions of a massive crime wave-to-come. What is sad is that no one has a solution, although politicians, moralizers--often the same people--are free with exhortations, blame and promises. Despite the downturn in crime rates in the 1990s, we can ask if America in twenty years or so will go the way of other crime ridden nations--Brazil, for example. In Rio de Janeiro, as the danger of crime became pervasive.

> ... anyone who could afford to began erecting fences around their buildings and homes. The very rich stopped walking in the streets and swimming in Rio's picture postcard perfect beaches, and turned away from government to provide for their needs--whether educating their children, solving crimes or even directing traffic. Steadily the middle- class and rich have turned to private services to create a parallel world that insulates them as the one around them disintegrates.[106]

The U.S. with its rise in the numbers of gated communities and private security guards seems partly on the way to Rio. But there is an essential difference; in Rio the poor live in suburban slums, the favelas. while the middle and upper- classes live in the city. The urban pattern in the U.S. is the exact opposite with the cities housing most of the poor while most of those who have the means put physical distance between themselves and the crime-ridden cities and impoverished inner suburbs. In effect, the middle and upper- class have moved to communities where they are the overwhelming majority and can control the schools and the police. Thus, the pattern of abandonment is reflected in geography. Rather than trying often in vain to crime-proof their city homes the affluent move far away from the free fire zone.

One additional reason for relocation is an often overlooked side effect of city crime--the impact on insurance rates which run moderately higher to as much as double suburban rate charges. *Consumer Reports* offers figures that demonstrate car insurance premiums in the nation's cities run from $100 to as much as $3,000 higher than in suburban areas.[107] The same kind of city penalty is imposed for home insurance. In short, taxes are higher, services poorer, crime is more threatening and the cost of protection greater.

Peter Stein © 1996 from The New Yorker Collection. All Rights Reserved. [1996 12 09 071 PST .HG Heaven]

What to do?

There is no consensus on the proper program mix to reduce the crime rates that are killing off city neighborhoods. There is the historic "patience approach" which relies on past experience to suggest that in time, the present predator population will diminish as they or their offspring move into the law abiding mainstream. Despite the claims that better policing is responsible for the reduction in crime since the 1990s, the across the board improvement in most cities with different policing methods suggest that patience--an apparent change in cultural patterns...is the prime ingredient. At present the "get tough" approach is in the ascendant. In this view, we cannot wait for city crime to fall off gradually but instead must reform the criminal justice system to provide for swift, certain and severe punishment. This is in sharp contrast to the alternative view popular in the sixties which focused on

> ...the "root causes" thesis, arguing that crime is the result of systematic disadvantage, prejudice and neglect. Proponents of the latter theory recommend job training programs, economic development, and increased social spending within poor communities. The implicit assumption of the first position maintains that crime must be contained before conditions in the community can improve; whereas the second assumes that community life must improve before crime can be controlled.[108]

A third approach is communitarian: high risk areas fed up with being victimized can effectively reduce crime rates if they are given help in training and organization. Among other case studies experience in Sandtown-Winchester, a Baltimore neighborhood offers encouragement for this view. A largely black high-crime area with a population of 10,000, this area had half the city's median income in 1989 with 68 per cent of its children living in poverty. With health insurance virtually nonexistent for much of the community, a sixth of the babies born to community mothers had low-birth weights.[109]

It is noteworthy that the predominant form of housing is the attached two and three story brick homes typical of Baltimore, not the high rise tenements characteristic of minority-populated slums in New York and other cities. But a quarter of the housing units were vacant and until demolished or reused, offer drug dealers a convenient base for their activities. Drugs are the area's curse: In 1990, 462 community residents were arrested for crimes related to drug abuse and 570 Sandtown residents were admitted to some type of drug treatment program.[110] (Remember the *total* population is only 10,000.)

On the surface the situation seemed hopeless but not so in reality. Thanks to an active community program which involved increased neighborhood policing, voter registration, work with medical centers to improve coverage, work with private and public funding sources to expand home ownership and renovate existing housing, cooperation with city sanitation and park services. The crime rate has decreased substantially: a 72 per cent drop in murders between 1987 and 1993, a 15.6 per cent reduction in violent crime.[111]

What does this mean for the future of troubled central cities? First, it is clear that community involvement is crucial and second, that much can be done even in an era of shrinking resources. The unanswered question is whether enough improvement can be achieved in time to rescue severely impacted neighborhoods and to alter the reality and perception of central cities generally as dangerous places. Realistically, we can hope for more and encourage more community action, we can applaud the successes--and still conclude that most middle- and upper-income people have concluded that it is prudent to put some distance between home, workplace and the concentrations of predators.

The Predatory One Percent

In terms of numbers, the problem does not seem insuperable. The troubled central city population responsible for most of the crime and other social problems is extremely small: only 17 per cent of the nation's poor, about 1.8 million families live in extreme poverty areas where at least 40 per cent of the population lives in poverty.[112] This is only about 5 million people out of a national total of over 270 million. How many professional criminals? The hard core repeat offenders are generally considered to be no more that 6 or 7 per cent of all criminals. The total hard core is perhaps only 100,000, not a huge army in a nation of 270 million.

Given the numbers, why the startling trend toward gated cities, walled enclaves and private police? By the early 1990s there were 1.5 million private security guards in the U.S., three times as many private cops as in public police forces. And the size of prison population had almost tripled since the early 1960s far outpacing the 25 per cent population growth. Gated communities reflect the real numbers and the perception of menacing criminal activity as uncontrolled and apparently uncontrollable.

In 1995 one in three black men in their twenties were under the control of the criminal justice system on any given day, either in prison or in jail or on probation or parole.[113]

Up to the late 1990s survey data for major cities offered no comfort. No less than 42 per cent of New Yorkers were crime victims in the early 1990s, 9 per cent mugged, 8 per cent burgled, and 22 per cent with cars that had been broken into.

Adam Walinsky points out that in ten years the U.S. has experienced 200,000 dead from violent criminals, over 100,000 of them murders committed by strangers. In contrast, ten years of war in Vietnam racked up only 58,000 dead.[114]

Crime rates are directly linked to fatherless households and the fact that U.S. illegitimacy rate has been rising rapidly is cause for concern. The rate for the black population has more than doubled since Senator Moynihan predicted trouble-to-come in the early 1960s. The rate was "only" 25 per cent when he viewed with alarm. Now, it is over 65 per cent for the black population as a whole and 90 per cent or more in hard-core slum areas. And the rate is up for the white population. It reached 22 per cent in the early 1990s. It's no wonder that one prediction that seems all too real and menacing is that murder rates will double in the next twenty years.[115] As Walinsky indicates, the future killers are now in diapers or in kindergarten. They will reach the deadly teens around 2010.

The odds against personal safety are staggering: The U.S. Justice Department estimates that 83 per cent of Americans will be victims of violent crime sometime in their lives.[116]

For a change, there has been good news on the crime front for three consecutive years. In 1994, the nation's overall crime rate fell by 2 per cent with the rate of violent crime dropping by a healthy 4 percent. But the joy may be short lived. John J. Dilulio, Professor of Politics and Public Affairs at Princeton University cautioned in Walinsky fashion: "this is also the lull before the crime storm." The reason for his concern is "40 million kids, 18 years old and under who are about to become teenagers, the biggest group of adolescents in a generation, and many of them "fatherless, godless and jobless.[117]

Louis Freeh, the FBI director has also warned of an upsurge in overall crime based on an increase in juvenile crime. This portends "future crime and violence at unprecedented levels."[118]

In short, we can probably take seriously Adam Walinsky's predictions for an even gloomier future. The handwriting is on the wall: a wave of remorseless youngsters armed with some of the thousands and thousands of handguns and semi-automatics floating through the slums.[119]

There is a fundamental problem of changing the culture that produces large amounts of law breaking. As we have suggested, the decline in a moral order that has reduced stable-married couples with children to an endangered species in slum areas. This is generating a breed of teenagers one criminologist views as "temporary sociopaths--impulsive and immature." Given easy access to guns and drugs, we have a prescription for a wave of violence without remorse in the early years of the next century.

It is known that rising crime rates afflict many nations--East and West; that in historic terms crime rates were worse in earlier eras in U.S. history, not only in urban slums, but also in part of the South and the Western frontier and that high rates of unemployment (as in the 1930s Depression) are not directly linked to criminal activity.

There is some evidence that many youngsters from crime-ridden areas are salvageable. The military has had considerable success in turning around sullen, unmotivated young people. As Tom Philpott reports

> Military leaders have begun speaking out about a "lack of values" among recruits--values like honesty, self-sacrifice, perseverance, instilled in previoU.S. generations by families and churches and schools. There's concern, too, that the worst of today's youth have no conscience, no sense of shame.
>
> Marine General Jack Klimp describes four types of youth who try to enlist today. The first hails from strong families with solid values. A second group lacks that grounding but, despite "a society that has declined tremendously in the last 25 to 30 years, has gotten through undamaged." No drugs. No arrests.
>
> The third type "are kids who have been damaged in some way" through domestic abuse, drugs, gang activity, or other criminal behavior. The proportion of these "slightly broken" youth in society is climbing. "We decide who among them we're going to take a

calculated risk with. Who will we waive and allow in?"

Finally Klimp describes a fourth group of applicants in a way that, perhaps, no previous recruiting chief has. This group, say Klimp, is "truly, honest-to-gosh, no-kidding evil. And our job is to keep them out."[120]

It is clear that slum crime has a devastating effect on the job prospects of youngsters who choose crime as an alternative to regular work and who may thereupon be killed (homicide is now the leading cause of death among young black males), maimed, or at best end up with little serious schooling and a criminal record. A reputation as a dangerous area discourages investors from creating jobs in or near slums while criminal activity in schools makes it hard for teachers to teach and other students to learn. There appears to be widespread breakdown in the moral value system that has created a generation of violent, amoral youths--so lacking empathy for their victims that they cannot even display the ritual signs of remorse at their trials that would ensure a lighter sentence.

Over the past quarter of a century crime has been converted from an issue conservatives used to win political victories to one which has been fervently embraced by both political parties. This is partly a matter of hard experience. Rehabilitative liberal approaches seem to have failed. The assumption that tended to focus sympathy on the criminal as a product of a catastrophic environment, the victim of an absent father, a neglectful mother and bad companions has been replaced by a consensus in which kindly sentiments are directed at crime victims and attitudes toward violent offenders have hardened. They are considered candidates for long term removal from mainstream society. Perhaps the major area of disagreement between liberals and conservatives is whether murderers should be given life imprisonment or be awarded a lethal injection to ensure permanent behavior modification.

In brief, by the mid-1990s the attitude toward most criminals has been transformed into a mixture of contempt and loathing, partly because of the growing callousness on the part of younger, violent miscreants. Frequently the attempt to portray them as innocent victims of a pernicious society has infuriated a broad spectrum of the public. The result is a backlash against offenders, their lawyers and a system that punishes them lightly.

In this context, distance is a rational response. A ten or twenty mile separation from potential predators is a powerful defense.

The perception that despite a downturn in crime statistics central cities are presently more dangerous than suburbs is borne out by the data. One study in the Seattle and Portland areas calculated the crime rate for central city residents as ten per 1,000, while the suburban rate was only a tenth of that level. However, the risk of injury from all causes was actually higher for suburban residents because of the greater incidence of traffic accidents.[121]

Residents of distant suburbs are four times as likely to be victims of injury-causing car accidents as central city people because they "commonly drive three times as much, and twice as fast, as urban dwellers.[122]

Is the awareness of highway risk likely to alter population migration from cities to suburbs and exurbs? Most certainly not. Crime is seen as a personal threat, one that can be minimized by relocating. Traffic accidents are viewed as unfortunate incidents that can be minimized by use of seat belts, avoiding alcohol before taking the wheel, by using air bags, by better defensive driving, and in short, by exercising care. Risks can be greatly diminished. In contrast, crime is seen as a loathsome, highly personal threat, while automobiles are seen as useful, controllable and, in a sense, generic risk. And there is the gated community in the city and suburb. An estimated three to four million people live in them and as we have seen, the number rises by the hour. Once again as in casual clothes, auto emissions and swimming pools, California leads the way with half a million gated residents out of a state population of 30 million. Following the leader, some residents of older neighborhoods are attempting to retrofit open communities with gated streets, frequent speed bumps and most of all, private police.

In passing, we might consider the heavily punitive national consensus in favor of mandatory heavy sentences and spartan conditions for offenders. Given the eventual return of most of the prison population to the street, the result is likely to be less than benign: less crime while the prisoners are incarcerated; vicious criminal behavior once they are eventually, released. To be honest, the prospects that criminals will be returned to the streets will be diminished by very long sentences: prisons, may be turning into expensive gerontocracies populated by sickly lifers and other long-term inmates.

The cheering news that crime rates decreased substantially in big cities since the early 1990s may reflect, in part, better, more aggressive policing--cracking down on minor behavioral crimes like squeegee extortion or rampant graffiti. More likely it may be a reflection of more basic changes. It has been suggested that many crack turf wars have been settled and the corpses no longer pile up, that there is a decline in the number of roving slum teenagers and longer prison sentences are keeping hard-core offenders off the street. There is also a tipping point theory in which crime is viewed as an epidemic and a variety of small measures eventually cumulate to reach a critical point at which there is either rapid improvement or a rapid slide into anarchic criminality.[123]

To be sure, public attitudes toward crime are matters of perception rather than hard data. For example, from 1981 to 1992 violent crime victimization rates fell 9 per cent and were a hair lower in 1992 than in 1973.yet a 1996 poll by the *Chicago Tribune* shows that only 7 per cent of Americans think violent crime in the nation has declined in the past five years. Asked why they think the nation has a bad crime problem, 76 percent cite what they have seen on TV or read in the news; 22 percent base their belief on personal experience.[124]

It would be heartening to believe that we have turned a corner in the crime rates that have been killing off city neighborhoods but two facts must be borne in mind: First, given a generation of media portrayals and personal experience it will take a decade or so of good news to persuade the middle-class that cities are once again safe and second the current wave of out-of-wedlock, dropouts, gang members is like to increase over the next generation.

Another Candidate: Rotten Schools

Beyond high crime rates and the family breakdown there are those who zero in on the poor quality of central city school systems as a major culprit in the plight of the central city slum population and the growing social and economic polarization of society. The modest improvement in automobile industry employment in Detroit in the mid-1990s resulted in help wanted ads for high school graduates. Personnel department experience offers a bleak picture of the gap between central city school graduates and their suburban counterparts.

Many products of Detroit schools barely learned to read and write in high school, here they must master the complexities of computers and geometric equations. Of people in the inner city, half don't even know what the Internet is. Half of them have never used a computer. Suburban kids are better prepared.[125]

Baltimore's public school system offers a good example of the formidable obstacles to substantial improvement. *Time* magazine reported that

> Declining enrollment and an accelerating middle-class flight distilled the city's school population to the point where today more than 70% of students qualify for a free lunch, a standard marker of poverty. Nearly 35% of the city's pupils are absent more than 20 days a year, triple the rate in suburban Baltimore County. Fewer than half the city's ninth-graders passed the Maryland Functional Test in mathematics, which measures only the most basic skills. In Baltimore County. 85% passed.[126]

The decline in enrollment and the problems in student achievement was accompanied by a doubling of school staff. (This is a pattern very much like that displayed by the District of Columbia.)

> As the city's schools declined, their employees thrived. In 1950, Baltimore's public schools employed 5,463 administrators, principals, teachers and other staff. By 1995 it employed 10,622, nearly twice as many, even though its enrollment had by then actually dropped 7%.[127]

The *Time* magazine report went on to identify the root cause, a dysfunctional management culture. This included depositions on

> ...the"dance of the lemons," a phrase education researchers use to describe the way school bureaucracies shuffle unproductive, even dangerous, employees from post to post.[128]

> Internal documents acquired by *Time* include the results of a performance evaluation of the system's principals as of June 1994. Of the 175 principals evaluated, none received an unsatisfactory rating-- despite the abysmal performance of the schools.[129]

Can the gap be closed by structural reform or special programs and courses? Concern over the black/white gap in computer literacy was underscored by the assertion that "the swift pace of high-tech advances will only drive a further wedge between these youngsters." U.S. Census data for 1993 indicated a gap in home access to computers. Pre-K to high school, only 15 percent of families with less than $15,000 annual income vs 74 per cent for families with $75,000 plus had such access. The white/black differential was not quite as wide: 43 per cent whites vs 16 per cent black.[130]

The question is whether lack of computer skills is a primary or secondary contributor to future income differentials. An alternative but probably more realistic view is that it is less significant than the fundamentals: literacy, numeracy, standard speech and work discipline. As Robert Samuelson suggests, the skills our worst (i.e., lowest paid) need most are "basic literacy and good work habits. With these in place, computer competence can be added to the mix as needed.[131]

Frank Levy seems to come down firmly on the side of the pessimists. He capsulizes the key dilemma of the central cities and the poor inner suburbs. The problem is the poor school system. As he sees it:

> Human capital is becoming an increasingly important determinant of earnings. In the US context of locally run schools, growing income stratification by place makes it harder for poor and working class children to acquire large amounts of human capital. The natural set of solutions to this problem involves intervention by higher levels of government. But stagnant earnings and living standards make persons increasingly suspicious of higher levels of government.[132]

Levy points out that neighborhood sorting by income has been accelerating, and in many city neighborhoods average teachers "are often overwhelmed by hostile student attitudes, so that classes settle for 'treaties' in which teachers make few demands in exchange for student compliance."[133]

As Levy suggests, central city decline is rooted in the highly decentralized structure of the U.S. public school system. He offers two solutions. The first is to permit good students in poor districts to segregate themselves through vouchers or special schools: Save some, leave the rest behind. The second is large scale federal and/or state intervention in central city schools that will entail a combination of more funding and imposition of higher standards.

The first alternative may be partly achievable even in an era of continuing political correctness when politicians speaking for the poor are hostile to approaches that smack of elitism and education triage. A more costly approach, intrusive intervention, is less and less likely in an era of donor fatigue and deep suspicions of government. There is a widening consensus, not confined only to conservatives, that losers deserve even less than the little they get; losers lose because their inherited and/or self-inflicted defects mean that they simply can't cut it: intervention doesn't work. In this atmosphere, programs targeted at young children may generate some support but money for damaged teenagers and young adults will probably not be forthcoming until the youngsters show up in the justice system. After that, there is plenty of money for prisons.

The other side of a poor school system is its major role in driving out middle- income parents with school-age children. Gregg Geil hits the mark squarely: the poor quality of schools is one of the chief causes of middle-class flight from central cities.

> Quite simply, no parents who can help it will allow their children to attend a school where students are merely implored not to drop out, rather than one where students are challenged to learn and excel. Further, concerned parents understand that once a school or school district has started its decline, no amount of money or program chances will reverse its slide. [134]

> Suburban sprawl is driven as much as anything else by families wanting to live where there is high-quality public education. Strong property values and low crime rates associated with the suburbs result largely from people choosing not to live where public schools are in decline. One need only look at real estate advertisements to realize that schools are the most important factor that potential homeowners consider in choosing a location. [135]

In Washington, D.C. as in other cities with increasing numbers of parents are sending their children to private schools often at considerable financial sacrifice. One mother who lives with her four children in a small house with barred windows.sends her children to Catholic elementary and high schools even though she is a Baptist. "I'm buying peace of mind," she said. [136]

Salvation from Guaranteed Employment?

Wilson sees the essential problem of the ghetto as a lack of jobs. Given his damning portrait of many slum area job seekers he offers a laundry list of measures needed to bridge the wide gap between black men and the mainstream workplace: higher standards, more money and better teachers for ghetto schools, a government WPA-style program to provide work experience and useful employment to supplement the private sector job markets. Also better linkages between school and workplace, government-funded daycare and job training centers to make potential employees job-ready. Given this kind of expensive commitment, all else will follow: fewer out-of-wedlock births, lower crime rates, livable cities.[137]

This prescription may indeed be valid, but the political obstacles to its adoption are formidable. Simply put, the tide is flowing in the other direction--less federal and state aid, a massive welfare reform that will place additional burdens on central cities and economic and demographic trends that seem increasingly centrifugal.

Elsewhere I have offered similar recommendations for a national full employment effort based in part on New Deal, Depression-era jobs programs.[138] As I suggested, the huge costs of such programs might be offset by useful work accomplished and social problems and costs averted. In this political climate, there are no takers.

Conclusion

Over the course of coming decades we can anticipate a mixture of bad news--"city crisis grows worse"--interspersed with cheery bulletins like Bob Herbert's column in *The New York Times* in late 1996 entitled "The Cities Climb Back."[139] Like the heartening stories that have blossomed periodically since the 1950s, Herbert focuses on signs of a turnaround, lower crime rates, improvements in the areas of employment and neighborhood revitalization and civic pride.[140]

The long-term trend is in the other direction. Employment and population levels have declined in most central cities. They are poorer because people with the requisite incomes and choices have tended to move out. True there are attractive, overlooked investment opportunities there. As Michael Porter suggests , central cities enjoy strategic location, possess significant internal market opportunities for home-grown entrepreneurs and contrary to myth, they have substantial human resources, willing workers.[141]

What's wrong with this picture? Why has this startling potential for private sector development been overlooked. Are investors timid, prejudiced or ignorant? Amy Glasmeier and Bennett Harrison provide part of the answer:

> To a greater extent than he realizes, many of Porter's themes have been sounded before. For three decades, community development activists have recognized that some of the poorest neighborhoods in our biggest cities are among the most well situated in terms of access.[142]

Unfortunately as the preceding text indicates, many central cities are seen as high tax, high crime, poorly governed areas; expensive, unpleasant, even dangerous places to live and work. Certainly, there are are attractive opportunities there for niche markets: for some retail outlets , for some revitalized manufacturing, for some advanced services, for some capital intensive warehousing that risks machinery, not people. Abandonment is not in prospect. More downsizing is. There seems to be little doubt that with a number of exceptions, the year 2000 census will reflect a continuation of the long term population trends and job losses that began almost half a century ago. Telecommunications technology will play a supporting role in this scenario. Big city offices are vulnerable to triage on the Detroit model, there will be vacancies with only modest successes in adaptive reuse.

Volunteers and conscripts will continue to move out and away. Partly because of a profound change in the public consensus regarding the dysfunctional poor. In the 1960s, there was empathy and compassion. In the 1990s there is a mixture of contempt , exasperation and frustration. The '60s saw the poverty program, the '90s saw a binge of prison building, harsh welfare reform and gated communities.

Based on past experience we can probably predict that
- there will be substantial improvement in selected cities from the impact of niche markets for young singles and hard working, law abiding immigrants. The question is whether they or their offspring will linger in the cities after they have made it into the mainstream or join the exodus to the suburbs;
- there is also some question as to whether a sizable fraction of the immigrant second generation will opt out of school and work in

- favor of hustling;
- there won't be much substantial new funding targeted to hardcore slum area people or places in the next decade;
- we would do better to forgo "place" programs directed toward "gilding the ghetto" and resuscitating downtown in favor of "people" programs aimed at improved education and skills for the salvageable and trainable to help them get jobs (and thereby buy or rent decent housing) either in city neighborhoods, or away in the suburbs. Certainly, we can encourage expansion of post high school scholarships for school achievers. Another promising option is restoring "tracking" for promising inner city students: funding to assist the motivated to move into the mainstream rather than frustrating, costly efforts to reach and teach everyone.

In the political context of the 1990s and the probable climate of the 2000-2010 decade many small steps may be feasible. Big plans and big programs are unlikely and big promises are a ticket to frustration.

In one sense, the world of telecommunications represents a unifying rather than a socially divisive technology. Television and movies provide a common experience, a shared universe in which the proverbial rubes and hicks have become part of the mainland, but at the same time there are "informational black holes" and "electronic ghettos" for the poor. The educated handicapped and the educated frail elderly share in the new informational technology. In a real sense, the era when city sophisticates mocked the bumpkins residing only forty-five minutes from Broadway is long past, but cultural hipness and flipness is no guarantee of a passport to the learning and earning mainstream.

Graham and Marvin point to major international trends all of which are operating against the fortunes of traditional city centers and by implication, the poorly-educated central city poor.

Routinised back office functions that provide services and support to the headquarters and top offices in the world cities are now beginning to utilize telematics to try and spread out to lower cost locations in the areas around the major global cities--and, increasingly, farther afield. They are therefore using telecommunications directly to access cheaper labour, services and property outside the high-cost global command centres at a rapidly increasing rate. This is associated with the reengineering and downsizing of centrally located firms.[143]

Graham and Marvin "cherry pick" among those cities that seem to be making it--New York, Paris and London and the troubled industrial centers.

> The main problem seems to be that telematics are associated with a reduction in the capability of many advanced industrial cities to deliver employment of the quality or quantity required to sustain their socioeconomic fabric. While this applies to all cities, it is especially the case in the old manufacturing centres which have a weak service base such as Detroit and Liverpool. Here, talk of an 'electronic requiem' for urban economies may, in some extreme cases, be almost justified.[144]

What does this do to the social fabric in general between the hierarchy of advantaged cities and disadvantaged cities and the well-off and the poor in all cities? Graham and Marvin are pessimistic.

> In a context where many western inner cities already suffer the multiple disadvantages deriving from high localised unemployment rates, the effects of these labour market changes threaten further to polarize the social fabric of cities. The corporate elites who remain in permanent employment seem likely to become more and more detached, both socially and spatially, from the casualised and flexible work forces and the structural unemployment that surround them.[145]

What one can argue is that this social fallout can be--is being--minimized by relocation to pleasant places that screen out direct contact with the casualties. The impact is thereby reduced to sending money; e.g., welfare checks to feed and pacify the badly off.

Graham and Marvin pose the critical questions for many central cities: do we need them? Do they have a future?

> ...the development of a whole range of telemediated services mean that physical presence may no longer be necessary for banking, shopping, education, health services and perhaps even work. The annihilation of distance and time constraints could undermine the very rationale for the existence of the city by dissolving the need for physical proximity. As Gottmann asks, will telecommunications lead to a 'renaissance' of the contemporary city or do they signal the 'dissolution' of the city? Put simply, do telecommunications reinforce the importance of cities as centres of production, government and consumption or instead do they provide a substitute

for face-to-face contact and dissolve the need for physical proximity?[146]

As Graham and Marvin point out, the directions that telecommunications technology will take us is not preordained. Like other technologies, they offer opportunities for both social reform and social fragmentation.

> The same technologies can be applied to empower and assist disadvantaged groups as well as to disenfranchise or exploit them. There are many examples of disabled people, women and ethnic minorities benefiting substantially from telematics. Telematics can be used to strengthen the public, local and civic dimensions of cities as well as to support social fragmentation and atomisation. They can help improve urban public transport.. They can assist in the search for sustainable models of urban development... And profoundly different political styles or urban government can all develop, each using telematics in particular ways to support their approach.[147]

It seems clear that in the context of the closing years of the twentieth century, the new technology will accelerate the tendencies that were very much in evidence before the first computer screen was removed from its packing case. This almost certainly includes a widening geographical and social chasm between many people in the cities and inner suburbs and the people with the wherewithal to flee the malign images and distasteful reality these places have come to represent.

1. Michael A. Stegman and Margery Austin Turner, "The Future of Urban America in the Global Economy," *APA Journal*, Spring 1996, 159-160.
2. Letter from C. Peter Magrath, National Association of State Land Grant Universities and Colleges, April 22, 1996, 2
3. Anthony Downs, Urban Problems and Prospects, (Chicago: Markham, 1973), 32. The citation refers to a *Daedalus* article published in 1968.
4. Quoted in Helen Winternitz, "Jobless Rate for Youth," *The Baltimore Evening Sun*, September, 28, 1979, 1
5. William Julius Wilson, *When Work Disappears*, (New York: Knopf:1996), 141.
6. Alexander Ganz, *Boston and the Flight to the Sunbelt*. (Boston: Boston Redevelopment Authority, September 1977), 19-20.
7. Wilson, op.cit., 117-136.
8. Malcolm Gladly, "Hiring Practices Undercut Inner-City Poverty Efforts," *The Washington Post*, March 10, 1996, 16.
9. Ibid.
1 0. Ibid
1 1. Bennett Harrison, *Urban Economic Development: Suburbanization, Minority Opportunity and the Condition of the Central City*. (Washington, DC: The Urban Institute, 1952-53.
1 2. Sources of Data: School Achievement Test Scores, Baltimore City Department of Education 1990, 91.;Baltimore City Health Department, 1988;Low birth weight babies, Baltimore City Health Department, 1988, 81.;Work Disabilities, 1989, U.S. Census Bureau, STF 3-A, 1990;Labor Force data, U.S. Census Bureau, STF 3-A, 1990.
1 3. Janet L. Levene, "Still Life with Tourists," *The New York Times*, May 25, 1996, 31.
1 4. Personal Interview, Marie Howland, Director, Canton Survey, May 29, 1996.
1 5. Malcolm Gladly, "Plant Closing Opened Door to a City's High Tech Boom," *The Washington Post*, May 30, 1996, 1216.
1 6. Jon Jeter, "Taking a Businesslike Approach to Revitalize Poor Neighborhoods," *The Washington Post*, October 28, 1997, A-3.
1 7. Gary Lee, "Breathing New Life into 'Brownfields," *The Washington Post*, March 11, 1996, A.4.
1 8. Office of Technology Assesment, *The Technology Reshaping of Metropolitan America*,(U.S. Government Printing Office, 1995), 227-228.
1 9. Personal Interview, Dr. Marie Howland, Director URSP, University of Maryland, April 3, 1997.
2 0. Kirk Johnson, "Washington Steps Back; and Cities Recover," *The New York Times*, November 16, 1997.
2 1. Alexander Garvin," Chapter 2. "Ingredients of Success," *The American City*, McGraw/Hill: New York, 1996.

22. David Rusk, "How We Promote Poverty," *The Washington Post*, May 18, 1997, C3
23. For a discussion of the controversy, see Melvin R. Levin, *Planning in Government*, "The Fair Shares Controversy." (Chicago: APA Press:,1987).
24. Jacob Weisberg, *In Defense of Government*, (New York: Scribner,1996),131.
25. See James W. Hughes and Robert E. Lang, *Targeting the Suburban Urbanities: Marketing Central City Housing*, Fannie Mae Foundation Annual Housing Conference, 1996.
26. Robert D. Atkinson, "Does Information Revolution Pose Threat to Nation's Center Cities," *The Baltimore Sun*, October.29, 1995,E-1.
27. David Rush, *Baltimore Unbound* (Baltimore: The Abell Foundation, 1996), 22.
28. Ibid, XVII, XVIII.
29. Ibid, 23-24.
30. Ibid, 68.
31. Rich Hampson, "Architecture's Withering Heights," *The Washington Post*, October 21, 1995, F-1.
32. Ibid,F-12.
33. James Howard Kunstler, *The Geography of Nowhere*, (New York: Simon and Schuster, 1992), 190.
34. Paul Gargaro, "GM Begins a Game of Musical Chairs," *The New York Times*, July 20, 1997, 26.
35. Camilo Jose Vergara, "Downtown Detroit: An American Acropolis. An Immodest Proposal," *Planning*, August 1995, 18-19.
36. See Deborah Epstein Popper and Frank J. Popper, "The Great Plains: From Dust to Dust," *The Best of Planning*, (Chicago: American Planning Association: , 1989).
37. James Bennet, "A Tribute to Ruin Irks Detroit," *The New York Times*, December 10, 1995, 22.
38. John Tierny, "You Could Look It Up", *The New York Times Magazine*, September 24, 1995, 27.
39. Blaine Harden, "In New York City, Incumbent's Flaws and Assets," *The Washington Post*, November 3, 1997, A-8.
40. *The Economist*, November.2- December. 1, 1995, 89.
41. Steven G. Craig and DiAndrew Austin, "New York's Million Missing Jobs," *City Journal*, Autumn.1997, 43-51.
42. Peter Passell, "Financial Capital? Yes. A Model Economy? No., *The New York Times*, October 19, 1997, Section, 15.
43. Judith Havemann, "Some Major Cities Stem Outflow of Population," The *Washington Post*, November 19, 1997, A-1, A3.
44. Kevin McQuaid, *The Baltimore Sun*, November 26, 1995, E.2.

45. Michael Pittas cited in Roger Lewis, "Converting Office Towers to Other uses Could Help to Revitalize Downtowns," *The Washington Post*, November 4, 1995, F.3.
46. Rick Hampson, "Shattered Dreams in Skyscraper City," *The Washington Post*, October 28, 1995, F-7.
47. Jonathan Hale, "Living High in Lower Manhattan? *Presentation*, September/October 1996, 22.
48. "Savannah Leverages:Private Investment For Successful Downtown Redevelopment," *US Mayor*, October 23,1995, Vol. 62, Issue 17, 8.
49. William J. Stern, "C'mon Governor Pataki, Lead", *City Journal*, Vol 5, No.4, Autumn,1995, 12-13.
50. Charles V. Bagli, "Betting on a Bigger, Better Hotel Business,"*The New York Times*, November 23, 1997
51. "Oriole Park at Camden Yards : Who Pays, Who Benefits and Who Loses," *The Neighborhood News of Urban Studies and Planning at College Park*, September 1994, 1.
52. Ibid, 3.
53. William Fulton, "The Cost of Being Minor League," *Governing*, June 1996, 82.
54. Janet L. Levene, op.cit., 32- 33.
55. New Orleans, like Louisiana in general, has a long history of electing colorful rascals. When I made the mistake of suggesting that compared to whining hypocritical Yankee grafters, local politicians provided entertainment value for their peculations, a New Orleans planner emphatically reminded me that this was an outsider's view: "You don't have to live here," she said.
56. Bruce Weber, "Cities are Fostering the Arts as a Way to Save Downtown," *The New York Times*, November 17, 1997, A-24.
57. Ibid.
58. Allen R. Myerson, " A Big Casino Wager that Hasn't Paid Off," *The New York Times*, June 2, 1996, 1, Section 3.
59. Ibid, 1.
60. Patricia Reed Scott, *The New York Time*s, June 1, 1996, 18.
61. Interview with Jac Smith, President, Urban Agricultural Network, November 17, 1997.
62. Paul Johnson, *The Birth of the Modern*, (New York: Harper and Collins, 1991), 208.
63. Office of Housing Research, *Housing Research News*, Vol 3, No.2, June 1995, 5.
64. See Melvin R. Levin, *Outside /Looking-In*, (Needham, MA: DAC Press, 1993)
65. Ibid.
66. Peter Passell, op.cit., Section 15.

67. "Immigrant Assistance," *The Economist*, March 29-April 4th, 1997, 28.
68. Thomas Sowell, *Migrations and Culture*, (New York: Basic Books, 1996), 48.
69. "The Global Apple," *The Economist*, March 16, 1996, 28.
70. Ibid.
71. De Tran, "Beautiful Nails are Path To American Dream," *San Jose Mercury News*, January.3, 1996, 6.
72. Thomas Sowell, op.cit, 48.
73. Norimitsuo Nishi, "Stabilizing LeFrak City," *The New York Times*, June 6, 1996, B1, B4.
74. Celia Dugger, "For Half a Million,This is Still the New World," *The New York Times*, January. 27, 1997, 27.
75. Thomas Sowell, *Ethnic America: A History*, (New York: Basic Books,1981).
76. Joel Millman, "The Caribbean Solution: Remaking Inner City Schools," *The Washington Post*, September 17, 1995, C5.
77. Richard Perez-Pena, "New York's Foreign-Born Population Increases," *The New York Times*, March 9, 1996, 25, 28.
78. "New York City's Newest," *The New York Times*, March 14, 1996, A-22.
79. Bernard Wysocki, Jr. "Moving In," *The Wall Street Journal*, Oct. 10, 1996, 1.
80. Edwin McDowell, "Hospitality is their Business, *The New York Times*, March 21, 1996, D1, D9.
81. Ibid, 6.
82. Jonathan Foreman, "Bombay on the Hudson, *City Journal*, Summer 1997, 14.
83. Heather MacDonald, "Gotham's Workforce Woes," *City Journal*, Summer 1997, 42.
84. Lou Cannon, "Southern California's Boom is Latino-Led, *The Washington Post*, July 12, 1997, A3.
85. See Lee Rainwater and William L. Yancey, *The Moynihan Report and The Politics of Controversy*, (Cambridge: The MIT Press: , 1967)
86. See Melvin R. Levin, "The Economic Development Districts," *Community and Regional Planning*, (New York: Praeger: 1969)
87. Vernon Loeb, "Reports Tell How D.C. Government Failings Affect Lives," *The Washington Post*, October 19, 1997, B-1.
88. Interview with Katherine Hams, D.C. School Administrator, November 14, 1997.
89. Ibid.
90. Tucker Carlson, "How the G.O.P Can Remake D.C. O.K.", *City Journal*, Autumn 1997, 61.
92. Alexander Garvin, op.cit. Chapter 2.

93. See Luigi Barzini, *From Caesar to the Mafia*, (Bantam: New York 1977).
94. Neil MacFarquhar, US Public Housing Getting Closer to the Ground, *The New York Times*, June 2, 1996, 1.
95. See Oscar Newman, *Defensible Space*, (New York: Collier, 1973) Chapters 2 and 7.
96. Rosabeth Moss Kanter, " AT &T Call home; In an Era of Mass Layoffs, Can Cities Fight Back?", *The Washington Post*, January. 14, 1996,.1
97. Ibid,
98. Barbara Stewart, "Confirmed by the Market," *The New York Times*, November 7, 1997, 40.
99. Cited in Bob Herbert, "The Cities Climb Back," *The New York Times*, October 4, 1996, 36.
100. William Rasperry, "Urban Comeback," *The Washington Post*, January 24, 1997, A23.
101. Roger Lewis, "In D.C. New Signs of Progress and Purpose," *The Washington Post*, January. 11, 1997, E-1, E-12
102. Editorial Notebook, Philadelphia's School Wars, *The New York Times*, April 6, 1991, 18.
103. Lewis Mumford, *Sketches from Life: The Autobiography of Lewis Mumford: The Early Years,* (New York:Dial Press: , 1982), 5.
104. Peter Hall, *Cities of Tomorrow*, Blackwell, Oxford,1988, 364
105. Robert E. Park, *The City: Suggestions for the Investigation of Human Behavior in the Urban Environment*, R.E. Park et al. 1925.
106. Diana Jean Schemo, "A Common Bond: Fear of Each Other", *The New York Times*, December. 24, 1995, E-10.
107. "Ratings, Auto Insurance," *Consumer Reports,* January. 1997, 15-17.
108. Raquel Wexler, "Urban Violence and Community Revitalization: The Case of Sandtown-Winchester," *The Journal of the Community and Regional Planning Program,* (Austin: The University of Texas, , Spring 1995) Vol 1, 7.
109. Ibid, 11.
110. Ibid, 12
111. Ibid, 14-15
112. *The Promise of Housing Mobility Programs*, (Washington, D.C.: The Urban Institute, Winter 1995-96) , Vol. 25., No. 3, 4.
113. Fox Butterfield, "Many Black Males Barred from Voting," *The New York Times,* January.30, 1997, A-12.
114 Adam Walinsky, "The Crisis of Public Order," *The Atlantic Monthly*, July 1995,. 52.
115. Ibid, 52.

1 1 6. Ibid, 45
1 1 7. Ibid.
1 1 8. Ibid.
1 1 9. Ibid, 45.
1 2 0. Tom Philpott, "The Wrong Stuff," Washingtonian, October 1997.
1 2 1. James Gerstenzang, "Suburban Traffic Tied to More Deaths Than Guns, Drugs in Cities," *The Washington Post,* April 15, 1996.
1 2 2. Ibid
1 2 3. See Malcolm Gladwell, "Department of Disputation: The Tipping Point.", *The New Yorker,* June 3, 1996, 32.-35.
1 2 4. "Things are Getting Better? Who Knew?" *US News and World Report,* Dec. 16, 1996, 32.
1 2 5. Robyn Meredith, "An Exit From the Inner City; Training for New Auto Jobs," *The New York Times*, April 21, 1996, 10.
1 2 6. *Time*, October 27, 1997, 90-91.
1 2 7. Ibid, 91.
1 2 8. Ibid
1 2 9. Ibid, 92.
1 3 0. "The Haves and the Have-Not," Newsweek, February. 27, 1995, 50.
1 3 1. Robert J. Samuelson, "The Myth of Cyber Inequality: Computers Are Not Causing Growing Wage Differences," *The New York Times*, October 25, 1997, 55.
1 3 2. Frank Levy, *"The Future Path and Consequences of the US Earning Gap,"* Federal Reserve Bank of New York, Economic Policy Review, January 1995, 35.
1 3 3. Ibid, 38.
1 3 4. Gregg Geil, "Viewpoint," *Planning,* May 1996, 42
1 3 5. "The Racial Makeup," *The Washington Post,* January . 25, 1997, c-1.
1 3 6. Anthony Faiola, "Stand-ins for an Out-of-Service District," *The Washington Post,* September. 20, 1996, F. 2.
1 3 7. Summarized from William Julius Wilson, op.cit.
1 3 8. See Melvin R. Levin, *Ending Unemployment,* (: College Park: University of Maryland, Urban Studies, 1982), Chapter 12.
1 3 9. Bob Herbert, "The Cities Climb Back," *The New York Times,* October 4, 1996, 36.
1 4 0. Ibid.
1 4 1. Michael E. Porter, "The Competitive Advantage of the Inner City," *CUPR Report*, Summer 1996, 4.
1 4 2. See Amy Glasmeier and Bennett Harrison, *Public-Sector Presence Essential to Inner City Survival,* (New Brunswick: Rutgers University Center for Urban Policy Research, 1992)

1 4 3. Graham and Marvin, op.cit.,146. 'Telematics' refers to services and infrastructures which link computer and digital media equipment over telecommunication links.
1 4 4. Graham and Marvin, op.cit. 165.
1 4 5. Ibid, 168.
1 4 6. Ibid, 318.
1 4 7. Ibid,. 319.

Chapter 2

Suburbs: Teleworking And The Winners Circle

Suburbanization has an ancient history. The wealthy have sought refuge from city noise, confusion and filth as far back as Sumerian Ur (2200 B.C.) in Phararonic Egypt and Imperial Rome. Havens for the well-off emerged in colonial America when the populations of New York, Boston and Philadelphia were less than 30,000.[1] The impetus has been the same across the millennia: green and pleasant areas for the villas of the well-off away from city squalor.

Some early suburbanization in the U.S. took place with the help of ferry service. Brooklyn Heights development began in 1819, initiated by one of Robert Fulton's backers for his steamboat venture.[2] Railroads opened up new possibilities. Railroad commuter suburbs were created in Westchester County, in Long Island outside New York City, along Philadelphia's main line, around Boston and Chicago well before the Civil War. By 1874, Chicago had 52 railroad commuter suburbs.[3]

Kenneth Jackson cites an 1871 advertisement for potential suburbanites in Louisville, KY promising an atmosphere that is "delightful, cool, bracing and invigorating. No malaria, coal soot, smoke, dust or factories."[4]

Part of the suburban movement was fueled by the same speculative fires that stoked the Western migration: the promise of a suburban home was viewed as a sound investment in real estate.

There is another factor that stimulated the flight from cities. Jackson points out that the threat of urban disorder has been a powerful force in stimulating the exodus. Rome's crime rate and London's thieves were helpful to suburban real estate sales. In the U.S., periodic riots and other signs of mass unrest were a useful sales tool. New York developers, for example, underscored the continuing menace of the mob violence that erupted with devastating results in the 1863 draft riots.

Sidney Brower reminds us, negative attitudes toward city living date back far into the nineteenth century. The suburbs assumed all of the virtues of the country, but suitably packaged for city people.[5]

Brower quotes Frederick Law Olmsted, the planner for New York City's Central Park and a major influence on landscape architecture and city planning, on the pernicious nature of city living.

> It is an established conclusion that the mere proximity of dwellings which characterizes all strictly urban neighborhoods, is a prolific source of morbid conditions of the body and mind, manifesting themselves chiefly in nervous feebleness or irritability and various functional derangements, relief or exemption from which can be obtained by removal to suburban districts. (Frederick Law Olmsted in his report to the Riverside Improvement Company, 1868).[6]

The distinguishing feature of most suburbanization throughout the millennia was its upscale bias: you had to be rich to move away. As Charles Glaab indicates, regardless of their personal sentiments toward city living, up to the end of the nineteenth century most people had no choice: suburbs were for the well-off while the working class had to live near their workplace.[7]

> Although the rural ideal was always influential in American life, only the very wealthy were able to maintain homes away from the heart of cities until the coming of the horse-car lines in the late

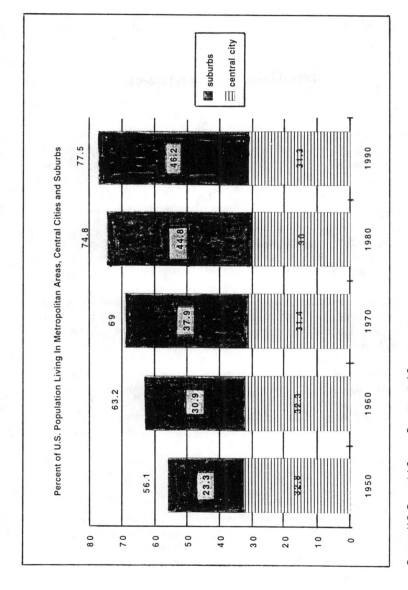

The vibrations were not confined to the moneyed upper classes. As Frederick Lewis Allen reminds us, as early as 1923 in a typical American city--the Lynd's "Middletown," half the working-class families owned automobiles.[12]

By the end of the 1920s the basic elements of America's urban pattern and problems were set in place. Most of the wealthy and much of the middle- class had moved out and by the 1950s they were joined by much of the working- class, including many city employees, teachers, policemen and firemen. The reality and perception of the central city as a place for losers, welfare cases and in general, poor, troubled people addicted to drink, drugs, sickness and crime was well on its way to becoming a fixed attitude long before the first computer was lifted from its carton.

As we move into a new century, the possibility of severing residences from city workplaces represents a geographic transformation that completes the distancing that began with the industrial city. In pre-industrial cities in the United States, Brower reminds us:

> ...where one *lived* was determined by where one worked rather than by one's social class. Merchants lived in the port area with carpenters, joiners, smiths, sailmakers, sailors, stevedores, and shipbuilders; tradesmen of all kinds lived interspersed with those who bought their goods and services; proprietors, merchants, and doctors lived with porters, laborers, carters, watchmen and criers. There was little clustering by ethnic group. Although there were wealthy sections of the city, these sections held a social mix because the middle classes required the proximity of less affluent neighbors such as tradesmen, servants, and artisans, and they supported the activities of beggars, prostitutes, and criminals. Many houses were subdivided for letting. The rich lived on the widest and best-paved streets, the middling families on the side streets, and the poor in the mews, alleys and courts.
>
> Social classes were mixed even within buildings. Since early times, the homes of the wealthy had shopkeepers and artisans living along the street fronts. In apartment buildings, the poor lived in the basement and climbed the stairs to the upper floors, and the wealthy lived in between. [13]

Kenneth Jackson details the powerful "pull" factors that stimulated the exodus from U.S. cities. Cheap suburban land, cheap gasoline, generous mortgage deductions, the practice of concentrating public housing in central cities rather than dispersing the poor to the suburbs.[14] But the "push" factors are equally important in triggering the outward movement. In Jackson's view:

> In comparison with the relative homogeneity of Denmark, Germany, England or Japan, the cities of the United States, and particularly the larger metropolises, have long been extraordinarily diverse in suburban terms. This has provided an extra incentive for persons to move away from their older domiciles----fear---millions of families moved out of the city "for the kids" and especially for the educational (as measured by standardized test scores) and social (as measured by family income) Civil rights activists and federal judges were unconcerned with the impact of racial change on cities and suburbs. Ordinary people, however, were concerned and because they loved and feared for their children, they simply speeded up a process that probably would have occurred a few decades later in any case...[15]

It would be a mistake to cast this vast demographic change in racial terms. The suburban migration includes black as well as white families and increasingly, older people and singles disenchanted with modern city life. The fact is suburbanization was well-developed in the U.S. before the Civil War when there were relatively few blacks in northern cities and in homogeneous England by the 1920s, long before the arrival of the West Indians and Pakistanis in the 1950s and 1960s. In part, this outward movement toward greenery suggests that a substantial part of the U.S. shares a common culture with the British that has survived two centuries after the traumatic divorce of the Revolutionary War.

By the 1920s, U.S. suburbs grew twice as fast in percentage terms as the cities. (Actual city decline did not begin until the 1950s.) Peter Hall points out that well before the end of Robert Moses' gigantic public works splurge, the 1930s motorists could commute thirty miles to Manhattan offices, three or four times the effective radius of the subway system.[16]

A combination of new roads, cheap mortgages and prolonged prosperity produced new corridors of accessibility from city centers to downtown offices and shops.[17] Zoning constraints ensured that most of the new suburban housing would be single family, built at a quarter of the density of central cities.

The rapid suburbanization of the postwar period was of course made possible by a long period of prosperity that permitted a great amount of public highway construction. Although there were no fundamental changes in transportation technology during the period, more powerful automobiles and more spacious highways extended the range of feasible commuting from that of the prewar period. Especially important was the interstate highway program authorized in 1956. The Act provided for a 42,500 mile, sixty billion dollar road network that, although not entirely complete by 1982, may well have been the largest public works program in human history.[18]

A Shortage of Good Neighborhoods

What the suburbs offer is an attraction that has become increasingly scarce in many cities: good neighborhoods. This means civil, non-threatening, compatible neighbors, not simply good housing. If housing alone were the criterion any number of cities large and small, could hold their own in price and quality with their suburban cousins. What the cities offer is a trade-off: put up with high crime rates, mediocre schools, high taxes and governments that have trouble removing the trash and clearing the snow in return for good housing in a central location. The suburbs focus on another trade-off, benefits and amenities with few disadvantages except a longer commute.

Suburban communities advertise, through the realty industry, the things that make them attractive: good schools, little crime, lots of greenery, good transportation access. Generally speaking, the objective is to make sales to people very much like current residents.

The critical factors in residential choice have remained stable, decade after decade. According to a 1973 study of Kansas City and Boston, most Americans want
- pleasant physical surroundings;
- no disruption from controversies over school integration;
- proximity to good shopping;
- a sense of safety from crime;

- good schools for families with children;
- respectable neighbors, preferably home owners of roughly the same income level.

Good medical facilities and quality of community services never arose as an issue. Apparently these were taken for granted. Single family, detached dwellings are a given; middle and upper income apartment neighborhoods were not included in the study. This omission does not affect its applicability to mainstream America where apartment living is a minority choice.

Asked in this study what they don't like and, in fact, what they are fleeing from, respondents gave answers which boiled down to an absence of lower class behavior. They didn't want

- gangs,
- crime,
- raucous night life,
- throwing trash outdoors,
- unkempt houses and yards,
- obvious signs of poverty, alcoholism, mental handicaps. [19]

What all classes are looking for in fact, is the obverse of the big city slum: They want a low density, green, clean neighborhood with good services and respectable neighbors.

The study suggests that twenty-five years ago, like today, in the choice of a place to live, the selection of a neighborhood comes first and the house second. This operates in the opposite direction. When a neighborhood is perceived as going downhill as a result of an increase in the crime rate, for example, the housing is usually in satisfactory shape but occupants are ready to move to a safer neighborhood where housing may be more expensive but the neighbors are more acceptable. Stability and predictability are crucial. Significant change is usually upsetting.

There Are Some Critics

Suburbs rich and poor have their critics, most of whom do not include the residents who have demonstrated their willingness to pay high prices for good schools, greenery and safety. In Sidney Brower's view:

Some feel that "the suburbs" have not lived up to their original promise (not rural enough) or their responsibilities (for conservation of land, efficient use of public utilities, equitable housing opportunities). Some feel that the suburbs are too built up, and others that they are too spread out. Some criticize the suburbs for being too uniform, too insistent on conformity, and some criticize them as unsuited to the needs of women, children, one-parent families, the elderly, and people who work at home; they foster social isolation and contribute to family distress and civic decay. Some find the suburbs boring and provincial, and some (a minority to be sure) prefer to live in the center city, or they prefer it at certain times of their lives.[20]

For the most part the criticism is muted. The near-hysterical diatribes against life in the suburbs so common in the 1950s and 1960s are quaint relics, abject lessons in media exploitation of public anxiety. One publication of the early 1960s is a sterling example. The subject headings in the *Split Level Trap* include the "Fraying Knot" (strains on marriages); "Grief in Paradise" (children and adolescents); "Absentee Father; "The Gimme Kids;" with side trips to the "Unhappy Singles," (Husbandless); "Chronic Losers" and other people whose emotional problems were allegedly linked to and exacerbated by living in the suburbs compared to life in the cities.[21]

The back cover of *The Split Level Trap* offers a tantalizing glimpse into behavior in a suburb (i.e., Bergen, New Jersey) that seems to be a hotbed of dysfunction.

Harried young housewives seeking escape in alcohol and stolen affairs. Husbands, frantic for success, staying later and later in the city. Teenagers, their every wish too easily gratified, steal and engage in open sexual promiscuity.[22]

By the mid-1990s most of the fears that life in the suburbs would create drastic and undesirable changes in behavior had long since dissipated. In fact, the sentiments against central cities had become so pronounced as to trigger a continuing and some cases, accelerating, exodus to the land of good schools. This school magnet is by no means a new phenomenon:

As early as the middle of the nineteenth century, Ralph Waldo Emerson had boasted of Concord: "We will make our schools such that no family which has a new move to choose can fail to be attracted either as to the one town in which the best education can be secured." In subsequent generations, such suburban school systems as Newton, New Trier, Scarsdale, Great Neck, Bethesda, Chappaqua, and Ladue became synonymous with educational excellence.[23]

Softening racial attitudes since the 1970s has altered some demographic perceptions for the better. Many affluent suburban neighborhoods have a significant number of Asians, some Latinos and a growing minority of blacks. The key seems to be the requisite level of incomes and a pattern of respectable behavior. The suburbanite's fear of duplicating the city's negative characteristics goes beyond people to include some types of buildings. In the author's experience, the introduction of high rise housing in single family neighborhoods is seen by many single family homeowners as a step toward "Manhattanization," a fundamental alteration in character in the direction of a high rise, high density environment injecting undesirable transients with few community ties or responsibilities.

If the apartment dwellers are renters they may be viewed as aloof from local concerns or else if condominium owners, they are seen as secessionists wrapped up in their building complex, swimming pool, shopping and services who also play no constructive part in town affairs. Even town houses are often seen as a threat to property values and neighborhood quality. It is for this reason that a proposed mix of housing types often meets with fierce resistance.

There is no question that perceptions of city dangers represents an increasingly important factor in suburbanization. Barbara Phillips offers scholars a crime-sensitive update of Ernest Burgess' concentric zone model of urban patterns developed in Chicago in the 1920s and 1930. In the Burgess configuration the poor live close to the city center and income and status increase with distance from the core. Based on post-1992 riot-torn Los Angeles a "decisive new factor" in understanding urban ecology is "fear," a perception of risk that is causing the rapid adoption of security strategies. These range from electronic scanning to gated cities as part of an emerging pattern of surveillance and physical barriers to repel intruders.[24]

The question is whether the blight of crime and fear will become so pervasive as to separate physically the victims and predators in the central cities from the refugee affluents who will include most of the middle- class and a sizable share of the working class. What is clear is that gated communities have spread from the West and Southwest to the East mostly in new subdivisions. Sometimes the phenomenon appears in unexpected places such as Greenbelt, Maryland, one of three New Deal era new towns which once prided itself on openness with a system of green pathways connecting homes to the civic center containing the elementary school, library, shopping and small offices. The community now contains a new fifty home gated community that charges a 20 per cent premium above comparable area housing for this apparent protection against a surge in local crime rates.[25]

Brower reminds us that gated communities come in two main varieties. The first is Serviced Apartment Houses in which the rental or maintenance fees for each apartment unit pay for shared in-house services, such as a doorman, swimming pool, lounge, and day-care center. In the United States, the model can be traced back to the residential hotels and apartment buildings of the turn of the century. Its present popularity came with the condominium movement, beginning in the late 1960s.

The second is Common Interest Developments which are based on a formal association of single-family houses, where mandatory membership fees are used to provide members with a range of residential services, such as security forces, nursing homes, and recreation centers. These may replace neighborhood services typically provided by public agencies. In the United States, the model dates back to the communitarian movement of the last half of the nineteenth century. It became popular again in the 1960s, with condominium developments.[26] (It may be noted that in the West and Southwest common interest developments may also include garden apartments.) As Brower sees it:

> Concern for security and tranquility is often behind the popular appear of this model. Residents see privatization as a way of gaining control over their residential environment. They see the neighborhood as a cocoon that surrounds them and protects them from the larger problems of society.[27]

Multiplex vs Blockbuster: Segmenting the Market.

A range of residential selections is available in most metropolitan areas. The "rose on the dunghill" (an Englishman's colorful description of an affluent suburb adjacent to a depressed New England textile city), is in reality a bouquet. For segmented communities, focused marketing is the name of the game. "Rifle shot" advertising, customed-tailored mailing to particular neighborhoods and advertisements in special interest publications have long been a powerful marketing tool. Its use in politics also has a long history. To cite one example, there is the Republican party mail and poster slogan of 1946 directed specifically to black Chicagoans that furthered that year's impressive Congressional gains: "They Kill You in the South and Kid You in the North." (This was shortly after the lynching of a fourteen year-old black boy from Chicago who, while visiting his Mississippi kin, was alleged to have whistled at a white girl).

It was not until the marketers led the way, followed by the political consultants that the social scientists demonstrated a serious interest in this kind of small area cultural analysis. True, in the 1930s W. Lloyd Warner had divided Newburyport (Yankee City) into six socioeconomic classes divided by income, reputation and location but it was not until the 1980s that Michael Weiss categorized, often with humor, a sophisticated subdivision by class and taste.

In one of his categorizations, life-style residential areas were identified by zip codes that are assigned to over a quarter of a million census tracts. They range from *Blue Blood Estates,* a category that parallels Warner's *Upper Upper* class; old money, family, reputation. The percentage of total households just over one percent is also close to Warner's figure.[28]

Down the ladder Weiss identifies other areas: *Money and Brains* (upscale enclaves of townhouses, apartments, and condos). *Furs and Station Wagons* (new money, corresponding to Warner's lower upper class), *Urban Gold Coast* (money, old and new in select city neighborhoods) and *Pools and Patios*, (older, upper middle- class neighborhoods. All told, these five high income life style clusters represented ten per cent of U.S. households in the 1980s. The increase in upper level income since then has probably added a percentage point or two.

Future demographic groupings spurred by the telecommunications revolution, will be determined in part by the substantial incomes enjoyed by these households as well as their orientation toward different area choices. Similar geographic clusters are present in most U.S. metropolitan areas. This orientation also changes with age: retirees may follow their children and grandchildren or choose their own growing subcategory in a golden age cluster, *Affluent Actives*. In short, while in their working years, affluents may be more or less loosely tied to a particular area, but as they gain influence and age, the linkages may either strengthen or weaken, depending on personal preference. To put it succinctly; these upper level strata are the very core of the intercommunity and interregional amenities competition based on rival area attractions.

The backbone of upscale suburbs is found in the 5.8 million households in the landed gentry cluster and the 7.8 million households in the *Elite Suburbs* category. Increasingly, they may also draw from groups that have in past years been attracted to big cities such as the *Artistic Bohemian mix* and *Money and Brains* category who may relocate after city neighborhoods are viewed as too risky even for the tolerant and the adventurous when key city services deteriorate to the point of total neglect. Moreover, affluent suburbs are beginning to sprout a modest range of bookstores, coffee houses and cultural events that make them an acceptable alternative to big city lures.

Hughes and Lang recommend that cities interested in attracting upscale residents to offset their concentrations of lower income people take aim at niche markets--part of the five urban uptowners group. Prime targets are the *Urban Gold Coast* (mostly singles), *Money and Brains* (childless, epicurean-oriented dual income couples) and *Bohemian Mix*, a diverse collection of highly educated singles. In the Washington, D.C. area, they point out that roughly half these upscale people now live in high density areas just outside the District and since they are mostly childless and adventurous, tending to perceive the city as an exciting place rather than as a gritty, dangerous, poorly serviced locality, they might be induced to move inside the city boundaries.[29]

The problems with engineering a sizable reverse flow in this group are formidable because they have made a rational decision to live outside the reach of DC government but still close to the bright lights. Moreover, if, District-like, mostly dysfunctional central cities continue to exasperate remaining residents-with-choices, the likelihood is that more will depart for nearby high density suburban refuges. The proliferation of luxury apartments in McLean and Falls Church, Virginia is echoed by similar developments in Chicago's Evanston and Oak Park, and Boston's Brookline and Newton, with examples in most other medium-sized to larger cities.

This trend predates the computer age, but as in the case of automobile-facilitated, family-oriented suburbanization, telecommunications technology offers a further means of severing or weakening ties to the central city. In short, retaining or attracting upscale residents is not solely a matter of repairing the public school system. A combination of experience with or fear of crime plus exposure to potholes, mislaid water bills, blizzards of parking tickets and a barrage of municipal scandals can repel even the most dedicated urban dweller. If this distaste can be assuaged by relocation to a nearby suburban citified area, upscale suburbs will continue to receive a sizable influx of singles and DINKS (double income no kids) persons, who don't make use of their good school systems, but like the combination of high quality services, low crime rates and proximity to their old central city haunts.

Commuting: A Minor Problem but Growing Worse

The attractions of the winners circle suburbs are such that they can afford to pick and choose among a steady stream of suitors. Approval is generally reserved for upscale apparel name stores, campus-type offices and research. Low-end convenience stores and fast food restaurants are muted and discreet; the Golden Arches are miniaturized, the 7-11 is screened by greenery. It is important to note that while commuting is not a frustrating, expensive time-draining experience in most well-off suburbs, there is a strong probability that the trip will get a lot more troublesome in the next decades. A Baltimore area survey, for example, foresees trouble ahead. In the mid-1990s just under half of freeways were rated as severely congested in the peak travel hour. By the year 2020, the proportion is expected to rise to nearly three quarters.[30]

In brief, in the Baltimore area as elsewhere, traffic conditions will get worse year-by-year and this reality will offer a strong incentive for people-with-choices--many of whom live in affluent suburbs--to get off the gridlocked roads. For most the misery is still to come.

To date most commuters take less than a half hour each way for the journey to work. There are, however, growing pockets of congestion of the kind that Pacific Bell uses in its California promotional films contrasting the happy, productive teleworker with the frustrated gridlocked commuter. The question is how soon we will actually witness the kind of large scale traffic paralysis portrayed in New Jersey:

> " New Jersey doesn't have rush hours anymore," jokes James Hughes, a professor of planning at Rutgers University in New Brunswick. "We have rush mornings and rush evenings punctuated by noontime backups." State planners estimate it will cost $13.8 billion over the next 20 years--close to double what will be available--simply to accommodate projected growth,let alone pay for the huge backlog of infrastructure needs.
>
> Along the traffic-clogged stretch of route 1 from the state capitol of Trenton to New Brunswick in the north, for example, transportation planners say some $750 million in improvement will be needed simply to keep traffic moving at today's sluggish pace. Without improvements, a trip that takes half an hour today could take over five hours by the turn of the century and in Morris County, the average commuter will be spending the equivalent of six weeks a year sitting in a car stuck in traffic. [31]

While there are occasional tie-ups, to date complaints concerning horrendous traffic congestion in the New York-New Jersey area seem to be more reflective of scholarly and media attention rather than the actual relative degree of road misery for most commuters. A 1996 survey shows that Los Angeles, Washington, D.C. and San Francisco-Oakland are the three leading areas for traffic congestion with two other California areas (San Diego and San Bernardino) in the top ten.[32]

Traffic projections point to worsening conditions: over the next twenty five years a 70 per cent projected increase in traffic and only a 25 per cent gain in highway capacity. For example, in metropolitan Washington, D.C. By the year 2025 Michael Weiss asks, "What's to keep us from Los Angeles style gridlock?"³³ Remedial measures for Washington as elsewhere include smart cars and smart highways to speed up traffic and bypass bottlenecks, more bikeways, more walking trails and metro-trains circling the Beltway. Business leaders want more roads and bridges.³⁴

These forecasts of misery-on-the-way don't seem to jibe with Joseph Coates prediction that the proportion of Washington area residents working at home will rise from 10 per cent in the late 1990s to a stratospheric (and probably unrealistic) 40 per cent in 2025.³⁵ A reduction in journey-to-work travel on this scale, or an even more modest 20 per cent, would create a substantial easing of highway traffic especially if much of the remaining travel can be rescheduled to off peak hours. In any event the prospect--and reality--of miserable commutes should provide a major incentive for professional-level suburban workers to substitute communications for travel. Along with home offices, the relatively few telework centers in satellite communities operating in the late 1990s will metastasize all over urban regions to provide an alternative to much highway commuting.

Upscale suburbs have other, less visible assets than good schools and relative safety. Many contain small pockets of industrial buildings, relics of the manufacturing dispersion of the 1880-1910 era. Built with solid craftsmanship, some still contain manufacturing enterprises or warehouse operations, but others offer profitable examples of adaptive reuse. Converted to office buildings replete with lawyers, accountants, financial service ventures, advertising agencies and other professional white-collar operations, rents increase from industrial levels of $2-$4 per square foot to $8 to $14 per square foot for the new tenants--most of whom live in the vicinity. A second source of local white-collar conversions is found in the dead or dying suburban commercial strip developments desperate for new capital and new clients. Some of these undistinguished rectangular boxes have proved susceptible to conversion from pet stores, tanning parlors and pizza carryout to attractive, lucrative offices.

In accordance with the axiom that "to those that have, more is given," it is the affluent suburbs (along with parts of exurbia) that are likely to be major beneficiaries of the trend toward increased teleworking. For this to be true, we can assume that many of the new enterprises in the coming generation will have their start in people's homes, garages, or small nearby office centers. These will be significant industrial incubators of the future. Not only will the upscale suburbs have the taxpayers, they will have the little businesses that will grow into healthy sources of revenue.

It is clear that for a variety of reasons it is the upscale suburb, (I have termed these "winner's circle" communities), relatively impervious to economic recessions because the affluent minority choose to live there. These communities are in the forefront of the new technology. Trends of the 1990s--the shift to self-employment, contract workers and home-to-office telecommuting are in full bloom in places where people have the money, connections and education to benefit most. What does the future have in store? Madison, Connecticut (pop.16,000, fifteen miles from New Haven) may point the way. This is an upscale suburb where the average house cost $240,000 in the mid-1990s. There are almost 1,100 concerns doing business in the community, many or perhaps most home based.

Given a labor force in the 9,000-10,000 range, this represents a sizable proportion of the population. It is noteworthy that in wealthy Greenwich, Connecticut, more than half the 1,100 building permits issued in 1995 involved additions on alterations for home offices.[36]

The Zoning Barrier

There is no obstacle standing in the way of earning income by writing or painting in your home. Many localities also permit a broad category of uses under the heading "customary home occupations." From a community's point of view there are concerns that someone will sneak a business into a single family zone that will 1) create parking problems by attracting employees or clients, 2) will create other problems like distracting noise, lights, smells or other nuisances associated with the workplace rather than the home.

Businesses depending on working often fall afoul of zoning regulations when they change from an unseen-from-the-outside home office common to many if not most residences to a visible enterprise with employees and/or obvious late night work and traffic.

In the late-1990's zoning by-laws have lagged behind economic reality. Law breaking with communications technology is still not as common as the illegal apartments and multiple room rentals that are found in millions of single family districts but the trend is clear. More people are doing it, more and more want to or will be forced to do it and communities will either have to revise zoning regulations, wink at violations (as they do with accessory "mother-in-law" apartments and roomers) or face the prospect of losing valuable taxpayers.

This raises another issue. If other businesses pay special municipal taxes why not home enterprises and computer offices? Even if they are virtually invisible, such activities might contribute a bit more to local revenues via higher property tax valuations reflecting the double duty of the residence office just as home based doctors, lawyers and dentists offices are supposed to do. Alternatively, there can be special business taxes large enough to help the city but small enough not to generate feelings of harassment that could lead to resentful, mobile people pulling up stakes in favor of a friendlier location.

The mid-level managerial and technical staff layoffs resulting from the reengineering and downsizing wave of the late 1980s and 1990s has generated considerable pressure for rezoning. Most of the home office workers are entrepreneurs rather than employees but in either case the likelihood is that basic changes in zoning are in the offing.

In the mid-1990s two-thirds of communities across the country have laws that regulate home occupations, typically limiting them to doctors, dentists and other licensed professionals who are believed to provide needed services to neighborhoods. But as more people work at home, suburban communities have come under increasing pressure to change their residential zoning codes. Many have concluded that their regulations are outdated, especially since sophisticated telecommunications equipment makes it possible for people to run all kinds of businesses from home without neighbors realizing it.[37]

One of the major incentives for self-employment assisted by telecommunications is the mid-career layoff. In every downturn desperate ex-employees try startup business as an alternative to prolonged spells of joblessness. The economic downturn in the early 1990's was one such stimulus. Data from the Washington area are revealing. In 1993 Loudoun County officials estimated that more than half the 6,600 businesses in the county were home based. In Fairfax County the number of full-time at home workers tripled between 1980 and 1990. In Montgomery County the number more than doubled.

Montgomery County's strategic plan emphasizes the emerging role of home-based businesses. Home business owners who don't alter the appearance of their homes or have many visitors no longer have to register the business with the county.[38] The question for the future is whether local governments will impose license fees or raise assessments on homes-with-offices or will be content to benefit from larger scale local expenditures by these ex-commuters.

Zoning in Teleworking

The diffusion of teleworking has lead to concerns that single family residential areas would be stealthily infiltrated by home businesses. Overtime, their character would be subtly altered when the Tudors, colonials, ranches and split levels would be a facade, barely concealing hives of commercial activity in dens, basements, attics and garages. This issue has been a subject of controversy in a number of suburbs.

The growing number of small businesses operating out of people's homes raises a number of questions, some of them troublesome. There is, for example, the high failure rate of small enterprises. A decade or two back the unwary were cozened with promises of big returns through telephone sales, growing mushrooms in the cellar, raising worms in the backyard and addressing envelopes. In the mid-1990s issues of computer magazines lured readers with promises of sizable incomes for persons willing to invest in a variety of computer-oriented businesses ranging from assisting taxpayers to cut their federal or local tax bills, helping businesses to reduce utility outlays, paring total mortgage payments with accelerated schedules or "writing" personalized children's books. Then there are the old standbys such as travel agent, purveyor of home products or beauty care products for the cosmetically challenged. These are all problems but the upscale entrepreneurs who run into trouble are likely to be far outnumbered by the successes based on skill, caution and keeping the day job until profitability is assured.

Planners: Pro Higher Density, More Teleworking

Planners are generally in favor of the spread of telecommuting because higher suburban housing and population density (including "infill") focuses new development in areas already served by roads, sewers and other infrastructure, rather than leapfrogging development to

the suburban fringe. Insofar as they have analyzed the impact of the new technology, they tend to favor increased use of teleworking because it matches the planners' criteria in favor of
- reducing traffic congestion,
- saving energy,
- reducing air pollution,
- offering more residential choices,
- helping housebound people (the elderly, care givers,
- the disabled) lead productive lives,
- cutting down on stressful commuting,
- helping communities to develop their economies without degrading the environment.

On balance most planners would probably agree with the recommendations to push the teleworking trend up a steeper gradient. After all, it will take a cutback of more than one or two automobiles out of a hundred workers not engaged in peak hour travel to make much of a dent in highway use or demand for parking, energy or impact on air pollution. Of course, if the if the self-employed are included we are already up to three to five in a hundred, enough to slow down the need for an extra highway lane on the beltway. Clearly, it would be prudent for any area confronted with proposals for expensive highway improvements to alleviate congestion to explore the possibility of reducing traffic volume by active efforts to encourage teleworking.

Carrot and stick are both possible. Carrot: loosening zoning restrictions on working at home, significant employer or tax subsidies for home office equipment and maintenance. Stick: congestion pricing to penalize peak hour drivers, substantial toll fees and parking charges. We can anticipate forceful employer action or mandates in order to cut costs on office and parking space by requiring employees work away from the head office part time or most of the time even in the absence of government encouragement

There is a growing need for mixed use zones that would mingle teleworking with facilities for child care along with personal services like restaurants and health clubs. In fact there could and should be incentives for new developments to provide a telecommunications infrastructure. It should be feasible to encourage more builders to construct houses which include spaces suitable for home offices. Neighborhood centers could be encouraged as an element of mixed-use developments in new subdivisions.

CATHY © Cathy Guisewite. Reprinted with permission of UNIVERSAL PRESS SYNDICATE. All rights reserved.

One question is whether metropolitan areas will need a wake-up call in the form of a disaster like the 1992 North Ridge "incident," the earthquake and its aftermath, that temporarily paralyzed highway traffic in the Los Angeles area. In other regions there may well be some major precipitating catastrophe--fire, explosion, mud slide, earthquake, blizzard, flood, or some other cause for a serious breakdown in automobile commuting. This will be followed by pleas from public officials for workers to work at home and for employers to accept and stimulate telecommuting. Into the breach steps a bellwether company-- Pacific Bell, or another vigorous local communications firm,offering simple packages to make telecommuting easy and inexpensive.

Another scenario is salami sclerosis in the form of creeping highway gridlock of the kind portrayed by James Hughes and by Pacific Bell in its promotional videos. Given the tend toward highway demand far outrunning road construction, there may come the day when one too many cars backs out of a driveway and the system grinds to a halt. When we approach near gridlock and commuting time in many areas approaches two hours a day, telecommuting and teleworking may represent the only available solution by creating a consensus that it is now time to replace substantial amounts of highway travel by major increases in teleworking.

Separate Worlds

One feature of suburbs in the 1990s, one that is likely to continue through the next decade or so, is their disconnectedness from the people and problems of the central cities. To a degree this reflects a kind of complacent, self-congratulatory Social Darwinism that suggests in a meritocracy like the U.S., each gets his just desserts. The disciplined and hard working get to live where it's nice: the misfits, alcoholics and failures find their natural habitat among their own kind in the slums. Moreover, there is little sentiment in favor of a "shoulder-to-shoulder we're all-in the same boat, my brothers" communitarianism. As Barbara Phillips sees it:

> In the past generation, it has become clear that suburban voters not only feel separate from city folk, they are also unwilling to fund public programs that could possibly turn cities around... I call it the "moating and malling of suburban America" or the

"Yes, you *can* run and hide" syndrome. Whatever it's called, it describes suburban anti-urbanism.[39]

When we add the fear factor to the equation we arrive at fragmentation and polarization on a scale unknown in earlier times, a chasm between the trapped and the refugees reflected in incomes, social status and prospects.

Most suburban communities have rejected or are planning to reject the six standard arguments advanced in favor of outright annexation or metropolitanization which merges suburban government with nearby central cities. Many urban scholars, most big city mayors and their allies in federal and state government call for closer suburban -- central city linkages These are under the six headings of (1) *Economic,* (2) *Fiscal,* (3) *Social,* (4) *Moral,* (5) *Identity and high culture,* and (6) *Regional response* to regional problems.

1. *One Economy.* It is suggested that the fortunes of the suburbs are inextricably linked to the well-being of central cities. The central cities still provide a major share of metropolitan jobs and they are crucial in attracting incoming firms. Moreover, the metropolitan labor force after the current demographic transition will require large inputs from central city labor pools.

2. *Fiscal.* The decline in the fiscal strength of central cities will impose increasing strains on suburban taxpayers to finance welfare, education and other costs.

3. *Social.* The growing concentration of poor, dysfunctional people in central cities represents a time bomb that is likely to spill over into the suburbs in the form of crime; e.g., carjackings, burglary and robbery) and in a growing political and social chasm between classes and races.

4. *Moral.* The miseries of the central cities are unacceptable and to a degree, preventable. Religious and moral convictions call for a positive answer to the biblical query: Am I my brother's keeper?

5. *Identity and High Culture.* It is the central cities that give metropolitan regions their distinctive identity for outsiders and it is the central cities where high culture--opera, major museums, symphonies, ballet, legitimate theater--are located. Suburbs may have mini-museums, chamber music, societies and dinner theater but these are pale imitations of the real, indispensable article.

6. *Spillovers.* Metropolitan areas confront many environmental land use and transportation challenges that can't be remedied by individual communities acting on their own. Spillover problems are beyond the political and fiscal capacity of cities and suburbs. They require a metropolitan political authority that can persuade and compel communities to compromise and act.

The suburban response to these arguments can also be subdivided under the same six headings --

1. *"We have our own, costly problems."* Suburbs are faced with expenditures for infrastructure to reduce traffic congestion, to improve the environment, to cope with growing numbers of poor and most important, to maintain good school systems; i.e., "charity begins at home; we have no surplus funds for trouble central cities."

2. *"They're a black hole."* It is alleged that many central cities are badly governed, subject to waste, fraud and abuse. By implication, their problems would be less serious if they used their resources more efficiently and more honestly and the corollary; until they do so it would be foolish to give them more money to squander.

3. *"Their real problems are their dysfunctional people."* The implication of this reasoning is that outside fiscal/governmental intervention is not likely to have much impact on the key problem, a deeply troubled population characterized by high crime rates, poor, school achievement, family disintegration, substance abuse and a weak work ethic.

In response to the argument that central city criminals will follow the money trail to the affluent suburbs, the upscale communities can counter with the facts: most of the spillover of city crime is found in the inner suburbs where population composition is similar to the cities. Prince George's county adjacent to Washington, D.C. is a good example. In 1996 Washington with a population of 600,000 had 396 murders, the metropolitan area's poorest suburb, Prince George's, with a population of over 700,000 had 142. In contrast, affluent Montgomery and Fairfax counties each with a population in the 700,000 range, and also abutting Washington, had separately less than fifty homicides in the same year. One caveat: Prince George's homicides are mostly in the arc of poor, drug-plagued neighborhoods along the eastern border with the District.[40]

4. *"Brookline had the right idea."* Brookline is a middle- and upper-middle-class community that juts into Boston. In 1874, Brookline voters defeated, an annexation proposal by a vote of 706 to 299.

> Voters were not turning down growth or development, but were expressing a determination to control the physical and social environment in which they lived.
>
> After Brookline spurned Boston, virtually every other Eastern and Middle Western city was rebuffed by wealthy and independent suburbs--Chicago by Oak Park and Evanston, Rochester by Brighton and Oakland by the rest of Alameda County.[41]

The question: Was Brookline right? And were the other suburbs correct when they that similarly opted for autonomy?

From the viewpoint of communities that view with alarm the troubles besetting adjacent central cities and take pride in efficient governments and good school systems, the answer is a resounding "yes." Miami's woes in the mid-1990s, a toxic blend of bad government and overwhelming social problems, have prompted three unincorporated, mostly affluent areas of Dade County to form their own cities (e.g., Key Biscayne) and others are lining up to do the same while some affluent sections of the city (e.g., Coconut Grove) are demanding either secession or abolishing the city as an entity by merging it with Dade County.[42]

Obviously none of the past history of rejection (and present calls for autonomy) would have been effective were the arguments for integration on economic or other grounds found to be valid. The opposite is true: Suburbs have flourished, decade after decade, as the central cities have floundered. Any call to place their prosperity, their schools and their police in the hands of or largely under the influence of central cities has consistently met with rejection. (On the other side of the coin many big city politicians would be reluctant to have their power base diluted as a small minority island in an ocean of suburban votes.)

5. *"We Support the Symphony with Our Subscription."* Residential location in a suburb does not necessarily mean a cessation in attendance at plays or concerts, or visits to museums. middle- and upper income city residents may be more regular in their support but the audience at

Lincoln Center, Broadway theaters and ballet performances ranges from partly to mostly suburban. For that matter, the relationship between high culture and cities can be frayed, distant or one of moderately mutual tolerance.

The question of identity is not especially troubling. Suburban residents have been commiserating with each other, swapping horror stories and denigrating central cities for many years. Further central city woes may be cause for additional embarrassment but not a call for massive rescue operations.

6. *"Functional Metropolitanism is the Way to Go."*

Regional responses to metropolitan problems that spill over from one jurisdiction to another don't require annexation or comprehensive metropolitan government. Some like air and water pollution are effectively countered by state action and federal pressures and funding. Other like transit systems can be addressed by creating a special authority to build and operate metro. In short, suburbs retain autonomy in the areas that matter most while ceding a limited amounting of freedom to a regional organization. What does all this mean? More of the same: suburban residents with a plate full of their own personal and community problems, congratulating themselves or having escaped the horrors of the central city are out in front in reaping the benefits from telecommunications technology.

Inner Suburbs

The example of Prince George's County's troubled communities adjacent to Washington, D.C. offer a clear example of the vast differences between hard-hit inner suburbs and the affluent suburbs that stand to gain the most from trends in teleworking. The high crime communities in Prince George's County are one example.

In general, the key problem of inner suburbs: that they have most of the ills besetting central cities without the government headquarters, office buildings, commerce and culture that offset the disadvantages. Mostly, run down residential areas, there is not much to build on for a turnaround. What is more, the inner suburb decay seems to be spreading. As Richard Wade concludes: "The phrase 'inner suburbs' surely will join 'inner city,' as a shorthand for a long list of urban ills..."[43]

As Herbert Muscamp sees it, the inner suburb problem is a lethal combination of physical and social factors.

> What planners call the first-ring suburb, the belt of single-family houses built between 1947 and 1977 around metropolitan cores, is fast wearing out. But some planners believe that the recycling of the first ring is the key to determining the way that Americans will live in the next 50 years.
> The social glue in these communities has also weakened. The population of the first ring is aging. The parents of the baby-boomers, the Depression Era generation that pioneered what the historian Kenneth T. Jackson called the "Crabgrass Frontier" are long retired. Houses are not only physically decrepit; their designs are out of date. As recently as 1978, 79 percent of the residents in the first-ring-suburbs of Minneapolis were members of one-job-two-parent nuclear families.
> It is not an accident, for example that the first ring is poor in mass transit and other services. In the postwar decades, the wife provided many of the services. She shopped, drove and helped organize community as well as domestic life. Her entrance into the work force deprived the suburbs of a key component of its basic infrastructure.[44]

The failure of candidates to address, or even grasp, the acuity of the suburban malaise explains, in turn, much of the populist rage that currently threatens the two-party status quo. America seems to be unraveling in its traditional moral center: the urban periphery. Indeed, the 1990 Census confirms that 35% of suburban cities have experienced significant declines in median household income since 1980. These downward income trends, in turn, track the catastrophic loss of several million jobs, amplified by corresponding declines in home values and fiscal resources.

As the *National Journal* tried to warn largely inattentive policy makers several years ago, "older working-class suburbs are starting to fall into the same abyss of disinvestment that their center cities did years ago.[45]

1. Kenneth T. Jackson, *Crabgrass Frontier*, (New York: Oxford University Press, 1985), 12-13.
2. op. cit., 31.
3. Richard C. Wade, "America's Cities Are (Mostly) Better Than Ever," in *The Making of Urban America*,,ed .Raymond A. Mohl (Wilmington, DE: SR Books,1988), 272.
4. Ibid, 70.
5. Ibid, 10.
6. Sidney Brower, *Good Neighborhoods*,(New York: Praeger , 1997), 7
7. Charles N. Glaab and A. Theodore Brown, *A History of Urban America*, 3rd ed. (New York: Macmillan, 1983).
8. Ibid, 163.
9. Ibid, 164.
10 .Ibid, 10.
11. Mark Sullivan, *Our Times*, ed. Dan Rather (New York: Scribner, 1996), 56, 390.
12. Frederick Lewis Allen, *Only Yesterday,* (New York: Harper and Row, 1931), 136.
13. Brower, op cit., 42.
14. Jackson, *op.cit*, 73
15. Ibid, 289-290.
16. Peter Hall, *Cities of Tomorrow* ,(New York: Blackwell, 1988), 276.
17. Ibid, 292.
18. Glaab, op.cit.,324.
19. Summarized from Joint Center for Urban Studies of MIT and Harvard University, *America's Housing Needs: 1970 to 1990,* (Cambridge, MA: December 1973), Chapter 5.
20. Brower,*op..cit.* ,113.
21. R.E. Gordon, K.K. Gordon and Max Gunther, *The Split-Level Trap*, (New York: Dell, 1962).
22. Ibid.
23. Jackson, op. cit, 147.
24 E. Barbara Phillips, *City Lights,* 2nd ed. (New York: Oxford, 1996), 163.
25. Interview with Celia Wilson, Planning Director, Greenbelt, MD, October 15, 1996.
26. Brower,op. cit.,69.
27. Ibid, 70
28. See Michael J. Weiss,*Latitudes and Attitudes*, (New York: Little Brown, 1994).

29. James W. Hughes and Robert E. Lang, "Targeting The Suburban Urbanites: Marketing Central City Housing," Fannie Mae Foundation Annual Housing Conference, 1996.
30. Baltimore Metropolitan Planning Organization, "Outlook 2020: Crafting a Transportation Agenda for the Baltimore Region," 4
30. In 1980 the average commuter took 21.7 minutes to get to work and in 1990 only 20 minutes.
31. Robert Guskind, "New Jersey says 'Enough,'" *The Best of Planning*, ed. Melvin R. Levin, (Chicago: American Planning Association, 1989), 208.
32. Alice Reid, "Area Traffic Stuck in a Costly Jam, " *The Washington Post*, December 10, 1996, 1
33. Michael J. Weiss, "The Way We'll Be," *Washingtonian*, January 1997, 62.
34. Ibid, 61
35. Ibid, 60
36. Jon Nordheimer, "You Work at Home: Does the Town Board Care?, *The New York Times*, July 14, 1996, Section 3, 1
37. Ibid, B6
38. Peter Pae, "In Area/Homes Businesses are Booming," *The Washington Post*, October 24, 1993, B 1,6
39. E. Barbara Phillips, op.cit., 334
40. Philip P. Pan and Robert E. Perre, " D.C/ Prince George Have This in Common: Too Many Homicides," *The Washington Post*, January 12, 1997, B-6
41. Jackson, *Crabgrass Frontier*, 149
42. Mireya Navarro, "Rich Areas in Miami Talk Secession," *The New York Times*, December 16, 1996, A-12
43. Richard C. Wade, "America's Cities are (Mostly) Better Than Ever," in Raymond a. Mohl ,ed., *The Making of Urban America*, (Wilmington: DE: (SR Books, 1988), 244
44. Herbert Muschamp, "Becoming Unstuck in the Suburbs, *The New York Times*, October 1997, WK 4.
45. Mike Davis, " Ozzie and Harriet in Hell, Decline of Inner Suburbs," *Harvard Design Magazine*, Winter/Spring 1997

Chapter 3

New Opportunity Beyond The Suburbs: Exurbs, College Towns And Recreation Areas

The first three chapters of the text described the changes in urban development accelerated by telecommunications technology. The second chapter depicted the lamentable problems of many central cities, problems to which telecommunications contributes by adding impetus to the outward movement of both affluent people and office employment from the central business districts. The third chapter was focused on the traditional suburbs, particularly the upscale communities that are already in a flourishing condition and are likely to benefit further from telecommunication.

This part of the text concentrates on the exurbs and other far-out communities such as recreation areas and college towns that are already growing and are due for another boost from this new technology. These outer suburbs, college towns and recreation communities all began their growth at least a generation before modern telecommunications were developed and all will enjoy an added growth impetus with the help of this technology.

Relocating to the periphery of greener grass and away from cities has been a traditional form of migration. Railroad commuting was mostly confined to the affluent. The automobile opened up distant suburbs to moderate income people. With the concurrent dispersion of jobs to the urban fringe, "regional commutesheds" are taking on amoeba-like forms, fanning out as much as 100 miles in places like Los Angeles, Houston and San Francisco.[1] We must also add New York and Chicago to the list.

Over the long term, centrifugal trends toward dispersion have been powerfully supported by developers in search of low-cost land and by farmers delighted with the prospect of substantial profits from land sales. In contrast, planners, conservationists and many budget-minded government officials view this tendency to low-density dispersion with alarm. They fear the loss of prime farmland which might, at some time in the future, lead to agricultural shortages and higher prices although this argument has become less common because per acre productivity has risen so rapidly as to render trepidation over a forthcoming dearth of produce suspect.

A more cogent argument is the sheer cost of urban scatteration-- roads, sewers, water lines, public services. One alternative is "infill,," providing incentives for developers to build on close-in sites already served by sewer, water and other infrastructure. However, developers are usually opposed to infill sites because they tend to be expensive, particularly if there is stiff competition for scarce land. Such sites are often relatively small and may be difficult to develop because of physical problems compared to the sizable tracts available in outlying areas. The infill alternative in central cities and inner suburbs is considered less desirable because good infrastructure does not compensate for bad neighbors. Frequently developers can identify perfectly good reasons why such sites have been left vacant. And naturally, farmers are unhappy with an infill emphasis since it entails the potential loss of bidders for their outlying property.

A second alternative is "cluster" development as opposed to dispersion. This is an attempt through zoning and infrastructure construction to direct the inevitable outward movement into urban concentrations rather than endless quarter acre to one acre tract developments. From the viewpoint of many single family communities, clustering multi-family apartments is a type of layout that is often considered excessively citified and probably lower class, a threat to tradition. The difficulty is cultural. The former mayor of Fresno, California made this clear. "We can't seem to associate quality of life with any development pattern except the traditional single family house with your fence and your yard."[2] Americans seem totally attached to their "dream house"--four bedrooms on a half acre lot.

Governments, on the other hand, feeling the fiscal pinch, see no alternative to intervention to achieve the objective well-stated by the Governor of Maryland in mid-1996 at a conference of the Maryland Municipal League. He warned that the state "will go bankrupt" trying to pay costs associated with urban sprawl.[3]

A laundry list of proposals were offered in the Governor's new initiative entitled "Neighborhood Conservation/Smart Growth."[4] Why the concern? Among the costs associated with sprawl, Montgomery County is building sixty schools to serve a growing population in its fringe areas while closing a similar number of schools in older neighborhoods that are losing population.[5]

One special attraction for the cluster approach is that despite lingering suspicions, it is often politically feasible. Developers often favor clustering because higher density adds value to sites. Clustering, for example, makes it possible to build on one corner of a difficult site, leaving the remaining ledge and wetland as an open space or a park attraction for prospective buyers.

Despite changing political attitudes, managing growth in metropolitan areas to counteract sprawl does not change the fundamentals of demographic trends. Infill, even if successful, is likely to capture only a small fraction of the tide moving inexorably to the outlying fringe. Most gains are taking place in towns within two or three hours driving distance from metropolitan areas. This is really exurban growth within the penumbra of the metropolis. "People want to live within range of major airports, and cultural or sporting action."[6]

The question is whether the surge of exurban growth will be fairly evenly dispersed in traditional subdivisions or whether it will cluster in sizable communities, new developments or expanded older settlements. Roger Stough sees the trend favoring the satellite city alternative over low-density development. As he sees it:

> Unlike in the past the new technology systems make possible a potential leapfrogging of development centers over the urban fringe to the intermediate and far periphery. Thus development may follow a spatial leapfrogging pattern with significant growth occurring in existing (satellite) centers as much as 50-75 miles out from the urban fringe. These distant exurban concentrations while often relatively

small (10,000 to 20,000 inhabitants) frequently have significant land and physical infrastructures including good transportation connectivity with other closer in regional centers. Because of this infrastructure and because of the locational freeing effect the new technologies are having it is likely that these exurban centers will become the primary growth nodes of metropolitan areas over the next decade or two.[7]

Centrifugal Movement

The Office of Technology Assessment discussed, at some length, the anti-sprawl arguments advanced by planners which tend to prove that spread-out development costs more in roads, other infrastructure and travel than compact development.[8] This seems to be a valid argument. A government-assisted in-fill policy to encourage development on underused central city sites and in close-in suburbs would be an exercise in investment efficiency. But what this rational cost-benefit argument overlooks is the human factor. People seem willing to pay a premium to put considerable distance between their homes and their businesses and the dangers they perceive from living and working in the city or even in an inner suburb. Often they can't afford housing in premium winner's circle communities. Cost-benefit points in one direction. Fear and loathing point in the other.

1. *Exurbs -- Farther Out Suburbs.* Some upscale communities located on the periphery of a metropolitan area date back to nineteenth century railroad expansion, others to post-1900 disillusionment with the perceived negative impact of democratization of the suburban population movement. Upscale people were unhappy with their new neighbors. As Brower sees it:

> With the streetcar, and then the electric street railway and the gasoline engine, middle income and working-class families were able to move to the suburbs and commute to the center. Much of the new development consisted not of isolated houses in the countryside, but of rows of monotonous houses on look-alike streets, and commerce and industry moved out to them. By 1909, the mid-nineteenth century suburbs were being abandoned for newer ones further out.

Wealthy residents found their arcadian settings threatened by these intrusions. One solution was to create planned communities within boundaries where the rural beauty of the countryside would be protected--and even enhanced. A typical garden community was a planned development of detached houses, with large yards, set well back from tree-lined streets.[9]

It is not these long-established communities that concern us. We are primarily interested in what has happened since the end of the Second World War.

Sometimes called exurbs, many of these postwar outlying communities are already experiencing substantial population growth; and in some cases, business expansion, partly because they offer many of the amenities of the close-in elite suburbs. The schools may not be as good and the public facilities are not as high in caliber but they offer quality housing and good neighborhoods at affordable prices. There is the risk, particularly in the more distant suburbs or exurbs of moving into areas where schools are really inferior and where crime rates are high thanks to a substratum of locals who seem to be preparing for a casting call for a road company version of *Deliverance*. Moreover, if the new housing is located in a genuinely rural area, the new migrants will discover the downside of picturesque agriculture in a rural setting: odiferous manure, farm machine noise, pesticides and semi-sober, enthusiastic hunters popping away at any ambulatory mammal.

In any event, exurban areas are one of the major beneficiaries of telecommunications technology. By extending the range of residential options beyond the daily commuting radius of the automobile, the technology permits greater dispersion, more residential choices.

Doing the Numbers

An estimated 60 million people live in the exurbs, the territory beyond the suburbs up to sixty to seventy miles from central cities. Between 1960 and 1990 exurban counties added more population than other types of counties, rising from just under 40 million to almost 59 million.[10]

These exurbs offer more greenery and cheaper housing than closer in communities. For the most part their residents if not working in the local service economy are white collar employees of suburban office centers and sometimes blue collar workers in suburban factories. Sixty mile-a-day round trip commutes are common but high speed highways keep the travel time down to thirty to fifty minutes each way.

The pronounced growth of exurban areas--the counties just beyond the metropolitan boundaries is reflected in an estimated net population migration of 1.1 million in just four years, 1990-1994. The mass movement includes an economic as well as a population shift from the northeast to "heavily white enclaves in central Florida, the Southern Appalachian hill country, the North Carolina Research Triangle and Atlanta." Idaho, Nevada, Arizona and Utah also have been big gainers.[11]

What does the future hold? A growing chasm in demographic patterns between net migration gainers and the rest of the nation. One prediction: by 2020 in twelve states, mostly in the Plains, upper New England and the Intermountain West--more than 80 per cent of youngsters under seventeen years of age will be white. In another twelve states (including California, Texas and most northeastern states), young whites will be a distinct minority.[12]

Is this growing importance of minorities a death rattle for California and Texas? Certainly not. In Hawaii whites are already in the minority and the state has prospered. Hawaii with a minority Caucasian population (35 percent) in 1990 is living proof that a demographic pattern far different from the mainland is no barrier to sustained livability and substantial economic growth. Asians are already a very large population at the University of California at Berkeley and California is thriving. It is absolutely certain that migration and immigration are not necessarily hostile to economic growth.

Why do people choose to live beyond the traditional suburbs? They exhibit in practice what Ebenezer Howard proposed for his English new towns, the best of town and country.

The research of Nelson and Sanchez on outer suburban population, compared to closer-in suburbanites found that in the mid-1980s:
- Urban home purchasers pay for desired rural attributes with longer commutes.
- Slightly more primary and secondary wage earners work at home relative to suburban wage earners
- Exurban home owners on the average have less than 10 percent larger houses compared to suburbanites (1,624 square. foot

vs 1,506 square foot.) but double the lot size (54,432 square foot vs 21,904 square foot).[13]

Nelson and Sanchez point out that Americans continue to share a "latent, pro-rural" lifestyle which views "single family homes on large lots as the ideal."[14]

> Basically, exurban households want a rural lifestyle but with all the advantages of urban opportunities.... They may not feel the need for social services offered in urban areas but they do not wish to be too far from them.... Small towns within commuting range of urban centers offer larger lots and larger homes than can be purchased farther-in ...and inconsequential levels of crime, pollution and congestion...and other urban discomforts. Exurban living also offers many social attractions, including a sense of community.[15]

Naturally, far-out exurbs are actively pursuing new economic development to provide new tax base. Creating exurban work centers for telecommuters represents an attractive development alternative for communities that run into a dead-end in efforts to lure manufacturing firms. On a cost-benefit basis the pluses are clear: none of the nasty side effects such as noise, vibration, smoke, heavy trucks, dangerous chemicals often associated with industry, but instead, quasi-college buildings, quiet, well-educated professionals healthy tax surpluses, light traffic, no drain on police and fire departments, and few complaints from residential neighbors.

2. *The College Town* is a second type of community which offers extraordinary attractions. This is not the university in the city, the community college or the small sectarian institution. It is the Palo Alto, Princeton, Chapel Hill, Madison, Ann Arbor model based on institutions with an outstanding technical base that have engendered significant population and economic growth in their immediate area. Schools and services are good, even a town or two beyond the ivy, and they offer a rich intellectual and social mix out of all proportion to their professorial-student population head count.

These communities are not only major beneficiaries of the new technology, they generate it. One byproduct is a redefinition of industrial base to zero in brainpower as the key to new growth rather than traditional industrial parks or shopping centers. To an even greater extent than is the case in suburbs and outer suburbs, the incubators for business and industry are homes, garages, local office buildings and other structures that offer low cost start up space for brainpower-based businesses.

It is an undeniable fact that Americans are better at building houses than building communities. Our housing space-per-head, our appliances and furnishings are second to none, but because attractive cities and towns are relatively scarce, we have bidding wars. Tourists go abroad to marvel at Paris, Rome and London, at the Italian hill towns and French villages and the poor but strikingly beautiful towns in the Greek Islands, at the green belts in Britain and the plazas in Spain and its former Latin-American colonies. In contrast, U.S. communities are bland, acceptable, but dull. Home design is one problem; community structure is another. Over the long pull we have been addicted to incrementalism, adding subdivision to subdivision, trying, and not succeeding, to create genuine communities with attractive centers. Zoning regulations, building government centers and providing minimal incentives are not enough to ensure that we build attractive communities.

In a few places this hands-off approach has worked: In college towns, bypassed New England villages, southern towns with courthouse squares. And there are a few cities that have adjusted to modern times, remaining vibrant; thriving big cities include New York (despite its well-publicized afflictions), Minneapolis, San Francisco, Boston, Portland and Seattle.

One of the results of attraction scarcity is price wars. Housing prices, especially in college communities like Palo Alto and Princeton and others on this short elite listing have been bid up to the near stratosphere and those lucky enough to buy in when prices were lower frequently exhibit the last settler syndrome: "Now that I'm here, let's build a moat to stop further growth." Boulder, Colorado is a prime example.

In 1976, Boulder became the first city in Colorado to put caps on housing construction. The result was a huge increase in commuting because no curbs were placed on commercial building. The second shoe dropped on election day 1995, when commercial construction was reined in. Golden and three West Denver suburbs have also voted on construction quotas.

The reason is simple. People attracted to a desirable community fear that fast growth will abrade and degrade the qualities that brought them there in the first place. To cite another Colorado example, the mid-1990s witnessed a battle over a proposal to strengthen and widen the runways at Aspen's airport to accommodate long-haul jets. Aspen has a year round population of 6,000 and local ski resort operators who have not seen much growth in twenty years want the new runways. Aspen's mayor summarized the rationale for opposition:

> "One of Aspen's greatest assets has been lack of accessibility," reflected Mr. Bennett, a Yale graduate who owns a French restaurant here.[16]

Jonathan Barrett underscores the attractions of small towns that are within driving distance from a university and other regional magnets and less than an hour from a well-served airport.[17] You don't actually have to live in a university town to enjoy most of its benefits. Half an hour to an hour away there are towns that are cheaper than Palo Alto or Princeton, but close enough to share the aura.

Why college towns? People with choices such as the professionals, information workers and entrepreneurs have everything: excellent schools, good medical facilities, high quality services, low crime rates and an attractive ambience of university attractions ranging from concerts and theater to bookstores, coffeehouses, and fine restaurants. In addition, there is the perennial injection of youth and vitality that is a feature of university life.

A major reason for the attractiveness of the elite college communities is their location outside the mainstream of late nineteenth and early twentieth century industrial and urban development. They were bypassed, making local go-getters quite unhappy at the time. In the 1890s a Princeton businessman lamented the slow pace of local economic expansion and assured his audience that with the proper go-ahead business spirit Princeton could become another Newark or Camden. Thankfully, it did not transform itself into what became two of New Jersey's most deeply troubled manufacturing centers.

The kind of growth that burgeoned around Princeton and Palo Alto and North Carolina's Research Triangle since the 1960s manifests an entirely different character. This is the mostly of the post-industrial era campus variety, white collar, research oriented, architecturally and socially compatible with nearby campuses.

Witold Rybczynski's take on attractive college towns focuses on what he calls college cities; places like Charlottesville, VA (University of Virginia); Raleigh-Durham-Chapel Hill (University of North Carolina, Duke and North Carolina State) and Santa Cruz-Watsonville, CA (University of California at Santa Cruz). He adds Madison (University of Wisconsin); Ann Arbor (University of .Michigan); San Luis Obispo; Eugene, OR; Provo-Orem, Utah; Austin, TX; Bloomington, IN; Boulder, CO; Iowa City, IA and Lawrence KS to this list along with Burlington, VT.

As distinct from a college *town* , in his view, a college *city* has brains-oriented industry and partly for this reason they are not small. In fact, most are in the 10,000 plus category [18]

One caveat: On Rybsczyinki's list of attractions is the relatively low cost of housing. While the U.S. average in mid-1995 was $143,000, the price he gives for Charlottesville was $151,000, Provo-Orem, $124,000, Burlington $149,000 Raleigh Durham-Chapel Hill $123,000-$183,000 and in Santa Cruz $227,000. Unfortunately supply and demand rules and there are perils of popularity. Home prices in Princeton start at the $300,000 plus range while a home in Palo Alto is $400,000 plus in the late 1990s.

We can accept the nation that college cities and the smaller college towns have much in common. The size and greater job potential of the former are matched by compensatory reality: small College towns like Princeton, Palo Alto and Chapel Hill are the centers of an encircling ring of communities that develop under their reputational umbrella. In this respect their total populations and job potential are far larger than would be suggested by their modest size.

What the college town scenario also points out is the power of sentiment as a key determinant in choosing where you locate your business and residence.

By the thousands, graduates reluctant to cut the umbilical cord, wait tables, clerk bookstores and work at other menial jobs in Cambridge, Berkeley, Chapel Hill and Ann Arbor. Thousands of professionals will stunt their careers by falling in love with an area, turning down promotions, if this means relocating to places they see as darkest Siberia, places like Detroit or Cleveland, St. Louis or Birmingham.

Coincidentally, the college town has an important price advantage over the typical semi-tropical beach community or the ski resort which compensates somewhat for higher home prices. It usually has excellent public schools and parents can look forward to teenagers spending part or all of their college years as low cost commuters, bypassing the expense of dormitory room and board.

Is it possible to create a college town by retrofitting? College Park, site of the University of Maryland, is examining the possibility of converting itself from a nondescript commuter suburb into something approaching a classic college community. Led by the University's School of Architecture, the preliminary draft of the city's comprehensive plan based some of its recommendations for downtown based on a study of Princeton and Chapel Hill. The report suggested some directions for College Park.

Needed:
- Unbroken building facades along the main street, Route 1,
- parking consolidated behind buildings,
- two-to-six story buildings, and
- development of a town square or green and other design guidelines based on findings in the model college towns.[19]

The prospects: fair to middling. Some progress has already been made but the gap between town and gown remains wide. The report points out that College Park's cultural base seems to be very different. For example, Princeton and Chapel Hill both supported four bookstores in the mid-1990s. In contrast there are only two in College Park, one of which sells only comic books.

3. The *Recreation Area* is a third type of magnet community. Many of these, particularly in seashore and mountain lake locations, have already experienced sizable population growth and the growth of attendant businesses that serve ever larger numbers of people. In the past, recreation areas have mostly served seasonal visitors or retirees. The new element in their growth is business and industry that no longer needs close geographical links with traditional urban markets.

Even before the proliferation of telecommunications technology, many people discovered that they could work year-round where they play, that they did not have to wait for retirement to enjoy a balmy climate or scenic surroundings. What the new technology does is make it easier to leave behind the cold, and the snow, the gray skies for sunny fun places.

The drawbacks: Quite often, lousy schools, a feeble intellectual life, substandard public services (try and find a good college or a decent library), distance from family and friends. Another problem in some areas: rising crime rates tied to some degree to conspicuous affluence amidst near third world indigenous people.

If there is one basic demographic trend in the U.S. that seems to be gathering momentum, it is the population loss, stability or slow growth in areas with cold, snow and ice to areas with sunshine, pools and beaches. Americans seem to welcome air conditioning to escape extreme heat while traveling between homes, businesses and other destinations, but seem to view extreme cold as a grievous imposition.

There is no shortage of recreational communities for potential buyers. As of 1993 one study estimates a total of almost 23,000 recreational subdivisions in the U.S.. Leading states: Florida, Texas, California and New Mexico. The fact is that many such developments are less than satisfactory, promising much in the way of amenities and environmental sensitivity but failing to deliver. [20]

One problem is that residential communities are often judged by lower bench marks by government regulators, partly because of the prospect of more tax base. It is assumed that residents may not live there full time or are transitional or at the very least elderly, sedentary and immobile. Consequently, there will be little burden on roads, police or public schools. For example, it is believed that cesspools rather than costly sewerage systems may suffice. Many recreational communities are shoddy, reflecting a widespread pattern of poor land use and inflicting substantial environmental damage. The moral: Look very, very carefully for on-the-ground reality, do not base locational decision alluring artist's fantasies of things to come.

Not much serious analysis has been devoted to the role of unfriendly weather conditions in accelerating urban decline. It does seem logical that places where you freeze or bake must have a lot going for them if they are to flourish in population growth and economic expansion. The Toronto, Minneapolis and Boston areas grow despite being on the frigid side. Miami, Phoenix and Houston are sizable case studies in beating the heat. Certainly, traditional technology helps. Air conditioning makes the South livable in summer, central heating renders cold climates habitable in winter. And heating and cooling have been extended to vast enclosed areas, most notably to regional shopping malls. These point the way to what we do best: find engineering solutions to natural challenges.

The old adage that "everyone complains about the weather but no one does anything about it," obviously is belied by the massive population shift from frigid to balmy areas. For example, Florida's population increased from just under 7 million in 1970 to 15 million by the mid-1990s. In contrast, Minnesota, land of ice fishing on 10,000 lakes, increased from 3.8 million in 1970 to not much over 4.5 million in the 1990s.

There may be other dangers in paradise, natural disasters for example. We should be reminded that a total of over 6,700 earthquakes struck California between 1980 and 1991 with the big one, the monster shaker from the San Andreas fault yet to come.[21] And between six and fourteen tropical storms and hurricanes reached the U.S. Coast each year between 1984 and 1994.[22] Most tore up parts of the Atlantic and Gulf coasts, and also devastating some Caribbean recreation areas in their path. Natural disasters may provide an unwelcome shock to older retirees.

> Retirees are a prize in the amenities wars because seniors increase the tax base but use less public services and facilities than non-seniors. They don't commit crimes. And they don't add to traffic congestion during the rush hours.[23]

Seniors have a special affinity for balmy retirement areas where the living is easy, taxes are low and the pace is slow. The market is sizable: five per cent of seniors retiring each year--400,000--retire out of state. Between 1985 and 1990, twelve states had net gains of at least $100 million from senior population transfer with Florida leading the parade with an estimated gain of $6.5 billion.[24] It is not surprising that by the late 1990s at least nine states and more than 100 cities and towns (mostly southern) were actively pursuing golden agers[25]

Often the move to warmer winters is accompanied by a fervor verging on the fanatic. There are woeful tales of past histories of constant earaches, chilblains, sniffles, head colds, influenza, troubled adenoids and sinuses and influenza--to say nothing of skids, dents, crashes in blizzards and ice storms. There are frightening stories of mighty snows, frozen automobiles, closed schools, dangerous treks to work place and supermarket, snow shoveling, snowplowing and just plain numbing weather too cold to venture out-of-doors. This is contrasted with one's healthy tanned spouse and children frolicking year around sans mufflers, overcoats, thermal underwear, galoshes and Goretex protection. As the Southwest retirement executive phrased it to the author, even if they have to stay where they work in the north, when they reach sixty, they poke their heads outside and say "who needs it?"

What can we do with modern engineering to help icebox areas? We can build mega malls like Minnesota's Mall of America.[26] This Mall contains four anchor department stores, seven junior department stores and 350 specialty shops. There are nightclubs, restaurants, bars and theaters, a LEGO Imagination Center and an eighteen-hole miniature golf course. In its center is a seven acre amusement park, Knott's Camp Snoopy. The cost at last count: $825 million in public and private funds. ($625 million private, $200 million public). Some other key statistics: almost 1,600 jobs, 30 million visitors in its first year of operation. There are almost 13,000 parking spaces and a monthly average of 130,000 transit riders to the mall.

What's missing? A proposal for 1.5 million square feet of office space and a convention center were excised from the original plan. An indoor lake vanished and hotel construction and an aquarium were postponed. And there are the usual suspects gone missing: government offices and high density residential construction. However a 9,000 square foot school serves high school juniors and seniors in specialized courses.

The hotels under construction may provide a partial response, but the prospect remains for a much more substantial recreational and residential construction within major climate-controlled space. The regional shopping mall, large as it is only a beginning. The super malls of the future will not only comprise a good deal of retailing but much much more. In time, we will probably see new recreation-oriented malled communities sprouting on the outskirts of large northern cities, followed in ten years or so by similar complexes on the far reaches of southern bakeoven-steambath cities. The obstacle is not lack of technological know-how or even a shortage of money. Imaginative pioneering is all that's necessary.

Clearly, the basic inventions in climate control involve heating and cooling small-to moderate-size places ranging from homes and automobiles to office buildings and shopping malls. Thus far Buckminster Fuller's proposal to dome sizable portions of major cities has not been adopted, although Milan's nineteenth century Galleria shows that it can be done. Instead, we have a successful example of an engineering solution to extreme cold in Montreal's miles of tunnels which offer the opportunity of ignoring blizzards and 30° below freezes by traveling, recreating, eating and in some cases, living and working in comfort, clad in suit jackets, not dressed like Nanook of the North. And we have air conditioned homes, offices and shopping malls in Miami and Houston which limit heat and misery to short dashes from one cool area to another.

The wave of the future is likely to be much, much bigger structures. It is the super super mall like the Edmonton, Alberta giant and the Mall of America in Bloomington, Minnesota. These two malls cover areas large enough to house a wide variety of shopping and recreation. What is missing so far is a climate controlled corridor to housing complexes and a range of offices and government facilities. A step or two more and we will have recreation and shopping-oriented new towns that will allow sizable populations to avoid temperature extremes.

Any self respecting economic development effort in troubled metropolitan areas can take action to
- shortstop expensive, distant recreation -oriented travel by providing a highly attractive alternative only an hour way,
- encourage employment growth by providing telecommuting services in an enticing, safe environment that offers either cocooning or interaction as desired and
- lure retirees, some of whom would prefer to remain in the area rather than pull up stakes for the land-without-snow.

In the summer of 1996 the Japanese National Tourist Organization invited foreign wholesalers, operators and travel planners to see Japan's imposing group of sixty theme parks. Of particular interest to Americans from cold regions is Seagaia, "a domed tropical beach" complete with sand and palm trees.[27] Why not in the U.S.?

The substantial balmy Mediterranean-climate growth of recreation areas in Florida and California and Hawaii suggests that climatic extremes can be avoided, but usually at a considerable premium. It also indicates that seashore, mountain, lakeside attractions now exercise a magnetic pull for the year-round resident rather than simply short-term recreational visitors.

Live-and-work where you play is a reality for millions piling into resort areas. Some are retirees linked to absent families by planes, telephone and marginally, by e-mail. Telecommunications barely figure into their decision to relocate. But increasingly it is active professionals and business people who are making the move and their operations are greatly facilitated by telecommunications.

The downside of living in many recreation areas is similar to the deficiencies of the exurbs: poor-to-mediocre schools, rising crime rates when native thugs and migratory criminals are attracted by affluent new comers. And frequently health facilities and services may be of far lower quality than in traditional older metropolitan areas. Finally, intellectual nourishment may be sparse. At best there may be a community college or marginal four-year institution within commuting distance. At worst, the new settlers, looking for signs of intellectual life beyond cable television, may be forced to lower their sights. Paradise has its drawbacks.

A Flood of Good Tidings

You would think that the exurbs experiencing a tidal wave of new development would be delighted with the prospect. Like the elite suburbs, one would expect them to be in a selective mode, picking and choosing among prospective suitors to cull out the drab, the unsightly and the environmentally threatening.

The possibility of attracting substantial numbers of upscale households and other quality development should be anticipated as an avenue to net tax benefits from high-end commerce and well-off people who either build sufficiently expensive housing to compensate for the school costs that represent the largest share of local public expenditures or who produce even more net revenue by enrolling their children in private schools. Furthermore, there is a dawning recognition that in the telecommunications era, home offices and garages are the new industrial incubators that generate business activity. In short, with a trickle to a flood of quality development heading their way exurbs should be making plans for long-term upgrading.

Not all of them are doing so: While their county planners are looking to preserve open space through such devices such as transfer of development rights (TDR) and on further developing existing communities, local farmers are loath to forgo the prospect of making a killing when the development market heats up. In their view, concentrating new urban growth around existing centers, many of them small, charming, and replete with architectural gems, does damage to the potential demand for outlying farm acreage. Similarly, local developers do not like to be told that dispersing housing on low-cost sites is often unsightly and always expensive in terms of increased road, sewer and water line construction and servicing costs. They much prefer to go on about their business in the traditional way: low density housing tracts and commercial strip development on local highways.

Where does the county political leadership come down in this controversy? About midway between the planners and the developers. They generally back preservation of open space but tough controls concentrating urban growth around small, attractive, existing towns to preserve substantial open space by public purchase are often too rich for their blood.

One reason is that they are impatient. Like most politicians they have a short time frame and waiting five years or more, staring at vacant sites, holding out for just the right development is a risky business. There is the constant pressure to bring in any new tax base, to accept any reasonable offer that brings instant gratification in the form of construction jobs and related activity. Exurbs are often filled with the not-so-well-off demanding tax relief and the most dazzling portrait of glad tidings for the steadfast and patient is no match for current urgencies.

The challenge facing county planners attempting to steer development away from exurban strip shopping malls surfaced in stark simplicity in Prince William County, fifty miles south of Washington. Developers and business owners charge that the county's ambitious attempts to control development and improve the "look" of Prince William and diversify the economy are unrealistic and may discourage

> ...the arrival of new retail stores--the meat and potatoes of today's commercial growth. In the latest skirmish, business leaders beat back a proposal last month that would have allowed officials to severely restrict commercial development along the county's major roads." [28]

Prince William, site of an aborted Disney company America Theme park, is part of a national debate going on in high-growth areas just outside metropolitan areas. As the former mayor of Pasadena, CA sees it:

> Prince William is part of a debate all over the country about what will work. Prince William developers argue that there is no point in waiting for office complexes that won't be built and that it's not in the county's best interest to leave parcels of land undeveloped waiting for projects that may never come. A lawyer, who represents developers says "you need tax dollars today. ...(Retail) may not be the golden goose but it is producing. I'm not sure that now is the time to be restricting the sector of the economy that is producing the vast number of jobs." [29]

What troubles developers? Efforts to beautify major roads and highways, by requiring more trees, limiting direct access from roads into shopping areas and severely restricting the size of retail signs. These proposals have been labeled as "anti-business."[30]

The political culture that generates sprawl and that focuses on short-term benefits is underscored by a *New York Times* series that stressed the prevalence of low-density development in western states, although the East is where it came from and where it remains in full bloom.

> "Nobody in this town has ever said 'no' to a developer" said Don Stenter, an air conditioner repairman and avid hiker who has been fighting the new developments on the northern edge of Phoenix. "We spend tax dollars to encourage sprawl, and then it comes back to us as air pollution..." (In Colorado) unable to raise property taxes beyond a certain ceiling, local governments are forced to get most of their money from as many retail outlets as they can attract.[31]

The missing ingredient in ensuring high quality regional development is a political culture that overrides the drum fire of attacks from landowners, developers and political leaders convinced that serious land use controls are a shortcut to disaster. Fortunately, there are a few exceptions to serve as models. For example, Portland is the living proof that assuring amenities is the path to desirable growth, a major weapon in the amenity wars that increasing determine locational choice.

> One corporate spokesman offered a pithy summary of the future with ambience becoming a key factor. "This is where we are headed worldwide," said Bill Calder, a spokesman for Intel, the computer chip manufacturer that has nearly 9,000 employees in Oregon.

"Companies that can locate anywhere they want will go where they can attract good people in good places."[32]

What did Portland do to deserve this accolade? It developed and implemented a regional plan over the strenuous objections of the usual suspects.

Ignoring lawsuits and pressure from commercial and development interests, Portland--acting on a much-fought-over new state law--simply drew a line around the metropolitan area, beginning in the late 1970's. On one side would be forests, farms and open space; on the other would be the city. The aim was to force jobs, homes and stores into a relatively compact area, served by light rail, buses and cars.[33]

The opponents of this kind of forceful action maintained the impossibility of containing a fast-growing city without turning away potential employers and depressing property values. The other component of the development plan seemed even more problematic: tearing up existing downtown freeways, limiting parking spaces and, in effect, force feeding public ridership on a light rail system. The results confounded the critics. Portland's gamble paid off handsomely.[34]

This is the key finding: Taking the long term view, holding fast to strong regional plans despite the critics, pays off.

...Instead of lost jobs, a silicon forest of high-tech campuses and factories grew inside the new urban boundaries. Instead of falling property values, home prices have soared. And instead of losing population, Oregon added 500,000 people, mostly in Portland and the Willamette Valley to the south, in the last 15 years.

It did so with the strictest laws against urban sprawl in the nation. Many of the newer companies in Oregon--among them Hewlett-Packard, Intel and Hyundai--say they moved here because there are forests, fruit orchards and meandering creeks just across the street from the contained urban areas. The employers said they wanted to locate in an area that could attract educated workers....[35]

Some time ago, the author was given the fundamental reason why institutional reforms are so often difficult to achieve. A state university system had purchased a troubled private college, pouring in a substantial sum without effecting any significant improvement. Why so little progress? An off-the-record comment by senior faculty gave the answer: "They made the mistake of giving the new money to the old men."

This observation explains the failure to reduce crime rates simply by enlarging police department budgets and to upgrade poor school systems by giving their current administrations the money they claim they need to guarantee positive results. So it is with many exurban areas. New possibilities for quality growth may be lost and opportunities foregone because power rests in the hands of people accustomed to the traditional structure. They are comfortable, more accustomed to mediocrity, fearful of leading the charge against resistant vested interests by making imaginative, politically hazardous decisions that require a long term view. The net result is likely to be more of the same; the undistinguished low density growth that characterizes many suburbs transferred to the exurbs. The only new ingredient is a proliferation of gated communities where private developers offer, at extra cost, amenity-rich autonomous environment: oases amid the prevailing blah.

One constant battle in exurbia is the struggle to convince locals that affordability is not necessarily the enemy of esthetics. This requires assurance to local merchants that a screen of trees and controls of signs and billboards can lead to long-term gains. Exposure to attractive mobile and manufacturing home developments in California, Florida, and the Southwest can sway locals who believe that low-cost housing is synonymous with debased pockets of impoverished fruit pickers' shacks.

Growing exurbs offer further room for conflict. In most exurbs, there is ample potential for cultural friction between newcomers and the old guard. If the newly arrived families are affluent and well-educated, they expect serious music lessons in the schools, tennis courts in the park, classes for the gifted, demands that strain the local tolerance for higher taxes. If the town goes the other way and fills its industrial park with low-wage industries, and lines its roads with minimum wage commercial establishments, it may find that it has imported some of the social troubles of the city along with the low-wage workers.

The Schools

One of the problems facing families relocating to exurbs, is the mediocre quality of many public school systems. One alternative (chosen by a surprising number of affluent residents) is to go private. High quality and very expensive private schools are enjoying a boom in much of the nation with thousands of parents dissatisfied with even the best of public schools, enrolling their one (or at most two children) in a form of education that combines quality, discipline, freedom from bullies, yuppie appeal, and parental expectations that their offspring will be in the winner's circle in an uncertain job market. This option is especially attractive to parents in so-so school systems.

There are other options for the hard pressed parent who may have more than two children. One is the academic track in not-so-good schools that have honors classes or some other academic alternative for promising kids, a quality oasis amid the mediocre to average teachers and student body.

A second option is home schooling. There are no accurate figures on the extent of home schooling, but Department of Education officials put the 1995 total at 500,000 children; one percent of U.S. school population.[36]

A third choice is upgrading the local educational system, a goal that may be feasible in some areas. If there is a concentration of new residents with substantially higher standards that the local norm, it may be possible to capture an elementary school or two with active parents riding herd on local school administrators and teachers. But while this takes care of K through 6, a turnaround of junior and senior high schools is much more difficult. The youngsters will have to mingle with the local kids when they reach junior high, considered by many as the most difficult years.

A fourth alternative is the charter school. Few ideas in American education today are building as much momentum as charter schools. The experiment that began quietly in Minnesota has taken root in nineteen states and the District and has spawned about 250 new schools. Scores of the schools, from California to Massachusetts, opened in 1997.

> Charter schools are a relatively new approach to public schooling. they are funded with public money, usually the same amount that a state or school district spends per student. But they operate independently of traditional public schools.[37]

Although laws vary from state to state it seems that almost any credible group or organization can develop an education plan and apply for a charter. Regulations differ in terms of oversight with some states requiring charter schools to subscribe to and uphold class schedules--but not pay scales. As a result, schools that receive per capita tuition funds from the states, and also require substantial parental involvement, seem to be able to achieve good results at reasonable cost.

Establishing a charter school may not be easy if there is strong opposition from the local educational establishment. For example, in Massachusetts charter schools have been stymied and delayed when space that was earlier available for lease was withdrawn because of public pressure. But despite obstacles, the movement prospers: the Commonwealth's first fourteen charter schools opened in 1995, and nine more opened in 1996.[38]

At the end of 1996, 500 schools were in operation and hundreds more are expected despite opposition from many local school boards and teachers' unions. Interestingly, although most proposals come from troubled cities, a number of school applications have been submitted by upper-income suburbs. Financing remains a problem even with public funds; the charter schools have to find their own buildings and usually some private backing to supplement public funding.[39]

Crime in the Country

There are other potential flaws to that ideal solution to the problems of urban living, relocation to an outlying suburb. Take crime for example. We have become so accustomed to identifying criminals as urban street thugs that we sometimes forget that rural areas breed their own distinctive underclass. Alcoholic school dropouts given to brawling, vandalism and robbery have lately turned to drugs and drug-related, often violent, crime. Living far out from an urban center does not necessarily mean safety or even low-crime rates. Motorcycle gangs, bored, drunken teenagers, drug dealing and assorted white trash can be a real hazard far, far from the city. But relocators may find that some exurbs harbor a home grown criminal class, with new recruits from drug-saturated high schools. Some suburbs are beginning to display the gang graffiti; turf-marking that presages crime-to-come. With this in mind, the move to the exurbs may represent an alternative to relocating to one of the suburban gated-communities springing up throughout the nation.

Perhaps the biggest shock in exurban communities is to confront pockets of inbred, morally retarded locals. Often dysfunctional, given to permanent welfare, spells in prison, hard-drinking and violent, they are prone to look on new residents and their homes and furnishings as delectable and defenseless--manna from heaven. New settlers who have grown accustomed to TV crime news as an endless parade of urban black and Latino thugs interspersed with an occasional luckless Mafioso are not pleased to find that WASP America has sired its own brand of true-blue, native-born underclass. Bonnies and Clydes and their descendants live on and on.

Whatever the choice of relocation, it pays to be careful to research one's prospective neighborhood. For example, some remote areas can boast of extraordinarily low indices of criminal misbehavior. Modest incomes are not necessarily a handicap. We can cite an advertisement for a manufactured home development near California's Yosemite Park, 140 miles from San Francisco:

- The crime in Tuolume county is practically non-existent;
- There were no murders in 1995;
- There are no gangs within a 50 mile radius.[40]

We might be skeptical of statement #1, pleased with #2 and inclined to research the accuracy of statement #3.

Depending on the extent and pace of cultural transformation, distant exurbs may offer medical challenges. None of us remain healthy all of the time and it is often a shock to discover just how backward many of these far out exurbs can be for a patient with anything but the simplest illnesses.

If there are medical complications, the sufferer will require the resources of a fully-equipped hospital that may be hard to reach quickly unless a helicopter can be made available.

What? Farmers Use Manure!

Newcomers thrilled at the prospect of living next to a picturesque working farm are often appalled by the discovery that fertilizer, pigs and pesticides can be smelly and perhaps more than an olfactory menace. Farmers also rise very early and their noisy equipment may disturb the slumber of homeowners not accustomed to beginning their workday at dawn. And there is always the possibility of an impatient line of commuters trapped behind slow moving farm machinery. Then there are the ubiquitous hunters who may combine nervousness, alcohol and eagerness. There is the instructive case of the New Hampshire lady who rashly ventured into her backyard wearing white knit mittens. Her slayer was acquitted by a jury of fellow huntsmen when he explained that at dusk at a distance of 100 yards, she looked like a deer in season.

Compromise is possible: Farmers and residential subdivisions can successfully coexist provided it's the right kind of farming. This means giving up cows, soybeans, corn and hay and shifting instead to tomatoes, peppers, lettuce, berries and herbs. This provides fresh produce for neighbors and markets. Locals can pick their own, buy from roadside food stands thus improving their diet and reducing their grocery bills.[41]

The alternative to mixing oil and water is to expand old communities to stimulate cluster development in high density new neo-traditional towns. This does not entail large public expenditures but rather a careful channeling of the stream of new investment in residential and commercial property that is heading out to fringe counties.

Secessionist America: Walled Communities

In the past two decades the number of gated-communities, developments with guards, walls, and controlled access has proliferated across the nation. Their hallmarks are "residents-and-guests-only" exercise centers, swimming pools, function rooms, tennis courts and grounds usually with small convenience stores, beauty parlors, laundromats and delis. It is appropriate to discuss this kind of development in this section because retrofitting existing subdivisions is difficult and often prohibitively costly and it is the open areas of the outer suburbs that are the locations for most new housing construction.

True, some close in suburbs are gated, particularly in the South and West, and some older suburbs and central cities offer a variety of controlled access in high-rise apartments, but it is in the open tracts some distance away from the troubled central cities and inner suburbs that new residents seem willing to pay a premium for added insurance against unwelcome intruders.

The growth of private residential governments whose latest manifestation is the gated-community can be seen in a few key statistics. There were less than 500 homeowner associations in 1962; 10,000 in 1970; 55,000 by 1980 and 150,000 by 1992. By 2000 there is likely to be a total of 225,000 with major concentrations in the Sunbelt. The nineteenth century privately-owned luxury subdivisions for the rich have been democratized to moderate income people.[42] This is more than Robert Reich's "secession of the rich." A substantial number of not-very-wealthy people are included.

In Southern California, real estate agents report that a third of all new developments in the 1990s have two features in common. They are gated and they are regulated by private "government," i.e., associations. Similar gated suburbs are found in quantity in Dallas, Phoenix, Washington, DC and parts of Florida. The advertising slogan for a Dallas enclave says it all: "Secluded from the world at large, yet close to all the finer things in life."[43]

The gated-community is, of course, far from a new development. It seems we have had the very, wealthy cordoning themselves off from the general population in much the same way forever. At the very least we have seen low density estate communities for the rich where the houses are invisible from the road, and where more often than not there are guard dogs and private security to deter burglars, solicitors and peddlers and to discourage casual drop-ins by outsiders. What is new is democratization, the extension of this option to the middle class and even some of the working class. Communities with walls or fences around the perimeter, private streets, sidewalks, parklands and recreational facilities and a supervised access controlling pedestrians and vehicles are increasingly available and popular.

Gated communities come in several varieties. "Lifestyle" communities, usually built for retirees or empty nesters, are located around natural open space, lakes, golf courses and country club-like recreational facilities. Suburban and exurban "new towns," providing affordability, recreational amenities and security, attract younger families. "Elite" communities offer distinction and prestige. These are enclaves for the affluent.

In "security zone" communities, fear of crime and outsiders is the key motivation for defensive fortification. Residents of such neighborhoods are concerned above all with personal safety, protection of physical property and maintenance of real estate values. Often these neighborhoods are within cities, and their occupants may be middle-class, blue-or white-collar workers or low-income housing tenants.[44]

One can argue with these definitions. Not all residents of the leisure world and similar communities for the elderly are especially affluent and neither are all the residents of the golf-centered or waterfront-oriented communities. They are certainly not poor: The purchase of $130,000 and up can come from the sale of one's old house and half one's salary and investment income plus the extras that can be derived from monthly Social Security payments.

The rationale for relocating to a gated community was eloquently stated by a Washington D.C. resident, the recent victim of a daylight car break-in.

> Like most urban residents, I'm not trying to avoid people of a different color or socioeconomic status. I'm trying to avoid the culture of crime and social decay I seem to find increasingly around me. Anybody who accepts basic social customs--don't rob, don't hit, don't urinate on the sidewalk--is welcome in my city. But when I pay high taxes to a city that can't pick up the garbage, maintain public civility, or keep my home, my car and my person safe from injury, living in a gated community starts looking mighty appealing.[45]

The author of this passionate explication is the executive vice-president of the conservative CATO Institute but his sentiments reflect a broader spectrum of opinion convinced that there is a disconnection between expensive city government and tangible, on-the-street results, most particularly in basic services and personal safety. For this reason minor signs of improvement--higher city school test scores, new downtown buildings, a fall off in crime rates--are not likely to sway most locational decisions. There is concern over physical secession of the affluent, city role models and potential leaders. The fear is that the U.S. is well on the road to a fearful Rio de Janeiro where the well-off take it for granted that they need costly barriers to be fenced-off from

CATHY © Cathy Guisewite. Reprinted with permission of UNIVERSAL PRESS SYNDICATE. All rights reserved.

predatory have-nots. These enclaves offer an affordable close-at-hand solution to pervasive fear of crime for a substantial proportion of the population.

Planners and others have underscored the negative impact of gated-communities: more Balkanization, more abandonment of central cities to the poor and troubled, secession of the well-off from those who have lost out. One stark prediction:

> Welcome to the new Middle Ages. We are building a kind of medieval landscape in which defensible, walled and gated towns dot the countryside. Will moats and drawbridges be included in the amenity package? Only ribbons of roadway, fiber-optic cables and digital electromagnetic signals interconnect these settlements. Even the commons where people might meet--the shopping mall or Walmart--is entirely private.

> Modern technology has accelerated the proliferation of gated communities. Sophisticated telecommunications, highway building and cheap gasoline, computer-based management and automated production allow longer commuting distances or no commuting at all. Proximity between workplace and residence is becoming less essential, as is the need for contact with other human beings.[46]

> Today one can shop, conduct business, engage in recreational activities, exchange ideas with people on-line or attend school without ever leaving home. And it all can be done solo. Face-to-face interactions are unnecessary. Thus, we are tending to build more and more private space.[47]

Graham and Marvin underscore another aspect of social segregation, closed circuit surveillance systems. As in other areas, the U.S. leads but we are not alone.

> The improved capabilities for integrating large numbers of sensors together--usually CCTV cameras--is leading to the development wide-area urban surveillance systems.

> This process reaches an extreme in the United States, where the extent and degree of ghettoisation and social segregation dwarfs that in most European cities...

The processes of privatising access to urban places, through post-modern architecture, private security guards, the building of controlled plazas and the walling of neighbourhoods, are supported by a sophisticated array of electronic monitoring and surveillance technologies--from computer communications systems in police departments to telematics-based alarm systems, infrared sensors, motion detectors and CCTV. Gated, walled, master-planned communities are increasingly common in the suburbs of all large American cities (and are increasingly on the agenda in the UK).[48]

Can Italy, Mexico, Spain and Russia--and other crime plagued, kidnapping-conscious nations be far behind?

Back To The Past: Going Neo-Traditional

An alternative to traditional subdivision development and gated-communities is the neo-traditional new town. These are high density settlements that offer many of the attractions of pre-automobile age villages.

Andres Duany's Seaside in Florida and Kentlands in Maryland and Peter Calthorpe's California ventures are two prominent examples of this development.

Peter Calthorpe's best known project to date is 800 acre Laguna West twelve miles south of Sacramento, California. The site is 800 acres (n.b. 640 acres = one square mile) and has five park-centered neighborhoods around a sixty-five acre artificial lake. The development focuses on a town center of schools, stores, civic buildings and tree-filled plazas along with low-rise apartments and houses. Houses are "zero lot lined;" i.e., no front yards and garages are located off back alleys where in James Kunstler's view is "where they belong" and everything is walkable.[49]

Other features favored by neo-traditionalists: back alleys for services and a mix of price ranges that include the town houses that have found disfavor with some exurbs because they allegedly attract relatively low-status buyers.

It's likely that the kind of clientele attracted to compact, pedestrian-oriented communities is self-selective. In effect, they serve what is probably a modest niche market that presently can't find safe, walking neighborhoods. Moreover, they offer an alternative to gated communities. In both approaches developers provide a secure, attractive controlled environment in places where local town and city planning and culture do not offer it. These are oases for people who cannot find-- or cannot afford--a desired level of amenities in the existing pattern of development in cities and traditional suburbs.

How big is the market for the neo-traditional community? One estimate by a Washington area builder who produces houses for the conventional and the neo-traditional markets:

> There is a niche market of home buyers who are willing to pay a 10 to 15 per cent premium to get a great lifestyle...most buyers are very young professionals without kids or above 50 years without kids.[50]

Sterility, Elitism and Fantasyland

A number of architectural critics and planners are extremely unhappy with the prospect of more "Disneyfication," the creation of controlled communities with a narrow range of architectural types, high prices to screen out at least half the population and an overall atmosphere of self-congratulation at having escaped the grunge and disorder typical of many long-established communities. Obviously, similar charges could be leveled at any upper income suburb and indeed, at parts of older cities.

In large measure, the criticism seems unjustified. Artifice made to appear natural and long standing is an old tradition. Central Park was a creation of Olmsted not an existing grassy, wooded area that was simply fenced in. Boston's Back Bay was only one of a series of landfills that began in the eighteenth century and continued through the twentieth. Over a twenty-year period the Back Bay was filled in with spoil from Needham nine miles away.[51]

Nantucket is only thirty miles off the coast of Massachusetts but nor'easters, chancy air services and erratic-in-winter ferry service can make it as distant as the north woods in bad weather. Nantucket has the kind of architectural charm--much of it new buildings that look old -- that makes many people want to extend their summer through fall, winter and next spring. True island romantics have, in my experience, gone to great lengths to put together a reasonable income to live there: raising rabbits, sewing baby quilts, trying to write or paint for a living, tutoring troubled school kids, running marginal mail-order businesses and falling into the old safety net--unemployment compensation. For serious professionals (aside from the rare writer who struck it rich), the problem of serving clients can be daunting. Some simply give up after being stranded once too often in an emergency. Others seek another New England paradise connected by all-weather roads to an interstate.

As for complaints about architectural style we need only remember that Art Deco, now treasured, was attacked as vulgar catering to the tasteless *nouveau riche*. Only much, much later did cast iron buildings and commercial structures once derided as garish (e.g., the Brown Derby, Lucy, the New Jersey shore's wooden elephant, Baltimore's Bromo-Seltzer Tower) come to be seen as humorous landmarks deserving of preservation. Victorian houses were once viewed as obsolete; the Eiffel Tower was once derided as a monstrosity--the list goes on and on.

The fact that the new communities are making strenuous efforts to screen out the squalor that afflicts older areas is nothing new; the country club suburbs have been doing that for many years. The difference is that the new democratized communities are reaching father down the economic ladder. As the Levittown migrant did a generation ago, the new suburbanites are taking advantage of the opportunity to distance themselves from the troubles and problems of older communities. This is one of the major sources of the disquiet. If the middle ranks of the urban population leave for controlled access suburbs, the cities and inner suburbs will become even more heavily concentrated repositories of the poor and dysfunctional. The question is what can be done about it, if, in a free market-oriented society, a sizable proportion of the affluent population chooses to relocate in suburban and exurban clusters? Public policy in the form of housing subsidies or fair share programs is not likely to make much of an impact on a powerful trend. The movement toward a segmented, income and cultural Balkanization seems unstoppable at least for the foreseeable future.

We can predict that a substantial minority of the population, especially people with school age children, are desperately looking for the Disneyfied, idealized communities of yore. And those that can afford it, perhaps 10 to 20 per cent of the U.S. total, will pay for a homogenized, artificial version of that past in neo-traditional or
traditional small towns, for housing in elite suburbs and for gated communities. Telecommunications will make it easier to live there

amidst greenery and often with privately-subsidized, quality services. When politicians speak of returning America to lost moralities they are not necessarily advocating relocation to controlled enclaves where the poor are effectively gated out, but that's what seems to be in the cards for the coming generation.

In looking ahead to this future we can learn some lessons from the past. One of the most important of these is the second generation's disillusion with the very qualities of intense community life and controlled, predictable neighborhoods that caused the first generation to settle there.

These neighborly communities are great for families with young children but stifling for non-conformists and confining for the teens. It's no accident that by their teens many youngsters developed an itch for bright city lights where excitement awaits and censorious neighbors and gossipers are far, far away.

These strictures apply equally well to the rediscovered small towns. Some of these communities were lucky enough to be bypassed by urban industrial grime, sometimes by chance, occasionally by choice.

Clearly gated-communities are likely to proliferate in outer suburbia. Another alternative is the Disney approach--a sizable community with a high quality school as principal magnet. A public school far better than the local competition apparently justifies a premium price for housing.

The Disney Company's new town in Orlando, named Celebration, is one manifestation of the heartfelt cry for a return to an idealized version of the past. It is fitting that the Walt Disney company is recreating a modern idealized version of a circa 1920s town. Celebration is designed to house 20,000 people. Although the professed objective is to provide opportunities for all income levels, the $650 monthly rental fee for apartments--the lowest price housing--will effectively limit minimum annual incomes to perhaps $30,000 since this is roughly the U.S. family median. It seems clear that half the nation's families would be excluded, unless some subsidy arrangement can be devised.

As *Time* reported in late 1995:

> Thousands of prospective home buyers recently converged on a former cattle pasture in Orlando, Florida hoping to become the first residents... The crowd piled into five tents where names were picked

from revolving bins...an occasional cheer went up as the winners were read out.[52]

What Celebration offers is clean streets and safety, no urban messiness, the promise of good schools, and first-rate health care facilities. In the context of the urban woes of the 1990s this is a "fantasy world so far removed from common experience that people are amazed at the prospect."[53] It comes at a price: buyers will pay 10% to 25% more for a Disney house than they'd pay for a house in a similar suburb, but they are getting a lot of extras for their money.[54] If Disney's projections are accurate, the company will invest $100 million on the project but stands to make at least three times that on lot sales alone.[55]

Buyers are given a choice of only six architectural styles with a price range of $120,000 for a townhouse up to $1 million estates at the high end. Since Celebration will also include modest rentals, the result, if all goes as planned, will be a mingling at the upper half of the U.S. income median. This is an unusually wide range in new suburban development.

Celebration owes a little to other models, traditional places much beloved by architects. The elevated green is inspired by Pisa Italy, the Plaza is Spanish style and the general layout owes much to East Hampton, Long Island.

Past models were only a starting point. As *The Economist* suggests:

> What people want in their homes, it seems are all the conveniences and technology of modern life, but hidden in "timeless" architecture...people also want a sense of community, hence the lakeside town centre, complete with town hall, post office, library deli, restaurants, bookstore-cafe, grocery, dry cleaner and needless to say, this being Disney, a 500 seat cinema. The town centre is meant to be a social place with people living above the shops and streets designed for strolling.[56]

Housing units "above the shops" is rare in U.S. suburbs. Overall, *The Economist* credits the Disney people with putting together successful ideas from elsewhere in its objective of creating both sense of place, lacking in other developments and variety of land uses.

The Economist asks a key question, " Is Celebration a revelation of how America will be?" Few companies have Disney's resources or the Disney image but as a Disney consultant puts it, with an estimated 80 million more Americans added to the population over the next 30 years, most of them in the South and West: "We need 3,000-4,000 Celebrations just to scratch the surface of their needs."[57] But can Celebrations be built sans Disney? One planner is optimistic that the afterglow will further the cause of Celebration style development.

> "Celebration is already making a big difference in getting people to accept this sort of development," says Walter Kulash who does transportation planning...and works on new urbanist communities all over the country. "We're well received when we say we're trying to do what Walt Disney is doing. The only thing I'm afraid of is that Celebration will suffer from the Seaside Syndrome. People will say, Disney can do it, but we can't."[58]

What do market studies say about the Celebratory approach to new communities? The 1996 national home buyers community preference vote, a survey of 1,650 buyers in Florida, California, Texas, Michigan and the state of Washington, points to a vote against typical all-sprawl and for dense town centers. But the four-to-one tally was in favor of a hybrid: cul-de-sacs, big yards, plenty of parking near stores and relative freedom in architectural choice.[59]

> Wanted: a town center with small shops and green space but with convenient parking; community gathering places; a mix of housing styles and densities but some neighborhoods with large, well-buffered lots and deep setbacks from the street.[60]

What this suggests is a heavy market plurality in favor of the kind of modified neo-traditional approach exemplified by Celebration. Buyer's comments respecting a wider choice of styles than the small six-course menu at Celebration could prove only a minor obstacle: the important features of the new community are not a matter of narrow choices but those of layout, focus and amenities.

The barrage of criticisms of the Disney approach to town development has already begun. What's wrong with the picture?

One critic is on the attack: Celebration is over controlled, elitist, undemocratic, it places too much emphasis on cars, it is banal, not Mayberry but Stepford.[61] Is this bad?

Ada Louise Huxtable, the architectural critic offers an apt commentary on the trend toward neo-traditional artificiality.

> Surrogate experience and synthetic settings have become the preferred American way of life. Environment is entertainment and artifice; it is the theme park with the enormously profitable real estate bottom line and a stunning record as the country's biggest growth industry...Distinctions are no longer made between the real and the false; the edge usually goes to the latter, as an improved version with defects, corrected--accessible and user friendly.[62]

John Henry points out that the housing will be priced well above houses in the surrounding area, a tribute to high Disney craftsmanship standards and also to the controlled sterility that he finds offensive. Interestingly he omits Disney's chief marketing feature, the quality of the new schools. As Randy Warner sees it the school is Disney's breakthrough. "Most of Celebration's buyers are probably sending their kids to private schools or are strongly considering it. When you balance $5,000 a year in tuition per child out of after-tax income against the premium for buying in Celebration you begin to see why they're lining up to get in."[63]

Reason number one for clones: if Levitt could net $1,000 profit on his $9,000 houses others tried to follow in his footsteps. If Disney can make a mint, deep pocket imitators won't be long in coming.

While we are waiting for more giants to come on stage, we have numerous examples of smaller scale developments that offer community amenities in contrast to customary subdivisions. There is a sizable amount of this kind of building in the Washington area's exurbs for example.

> Loudoun is the Washington area's fastest-growing county, and that growth is coming mostly in the super-size subdivisions of Cascades, South Riding, Ashburn Farm and Ashburn Village, which have sprung up east of Leesburg since the late 1980s.
>
> People living in the 11,000 homes in these new "towns" say they like the community spirit, the sense of safety for their children, the newness of the houses and the schools. There is always a place to

park and a new place to shop or eat in one of the dozens of strip malls along Route 7. They moved here because they could afford far more house than they could in Fairfax County, Montgomery County or other close-to-the-District communities.[64]

What these exurban communities offer thirty-five miles from Washington is a space and sense of orderliness much like traditional towns with

> ...village offices, village managers and work forces that maintain roads and take care of community pools, parks, trails and man-made lakes, and in some places pick up the trash. There are elections for officers in the home-owners associations, which function something like a town council, mediating disputes and enforcing strict rules governing everything from paint colors to mailbox location. To pay for it all, there are taxes in the form of association dues.[65]

The trade-offs for some residents are formidable--up to a thirty or forty-five minute commute. But with housing selling at a reputed half the cost of similar homes in Fairfax County, (a winners circle closer-to-Washington suburb), the residents seem happy with the choice of location.

What can we anticipate over the next twenty plus years? I would venture four predictions:

- Entrance of deep pocket investors, major firms in housing, insurance and building materials, pension funds and other large scale investors in the "Son(s) of Celebration II, III" sweepstakes. Many of these will offer half a loaf, like the Kansas City development I lived in that lacked the village green, playgrounds and much else available in the original template--Levittown, Long Island.
- Formation of architectural-engineering-planning teams that have or claim to have the Magic Kingdom touch in fashioning saleable-at-a-premium new communities.
- The new communities are likely to reflect the temper of the times in one respect: unlike Celebration there will be walls or other serious boundaries and in many, gates manned by private security forces.
- More house designs that include spaces suitable for home offices,

and mixed-use developments. Residents would then have the option to not work at home, but be within walking or biking distance of the work site. Thoughtful design would be needed to successfully integrate telecenters into the neighborhood. Regional telecenters drawing telecommuters from larger areas are also in the cards.

In the early 1990s telecommuting centers were off to a slow start. There were twenty to thirty offices in Japan, only one in Ontario, California which is sixty-two miles east of Los Angeles and preliminary plans for metropolitan Washington, rural Kentucky and Long Island, New York.[66] But the pace is likely to pick up by the end of the century.

Incubation-in-The-Den: Home Offices

It is always a sign of the times when the far out, the rarity becomes the commonplace, marketed like soap or soft drinks. So it is with home offices. Builders are doctoring their marketing to attract self-employed workers, people who are contemplating working at home and those who want at-home work space and extra telephone lines for computers.[67]

The cost? Ten foot by ten or twelve by twelve "bump outs" or finished basements for offices add $8,500 to $11,000 to the cost of a townhouse and almost $15,000 to the cost of a new single family home. As one Washington-area building company executive put it:

> With all the corporate downsizing and government RIFs (reductions in force), home-base consultants will need space and one spouse in a dual-income couple will want to spend more time working at home.[68]

There is also the time vs distance factor. Outer suburbs are usually no more than ten or fifteen miles away from large shopping centers or "big boxes" (e.g., Walmart). A half-hour trip is seen as a reasonable price for other amenities since any kind of metropolitan travel suburb-to-city or close-in suburb-to-suburb takes at least that long. In the Washington, D.C. area (as in other metropolitan areas), the big gainer is the outer suburbs where trip volume tripled compared to a relatively small 28 per cent rise in the center city.

Interestingly, average travel time to and from work has remained stable in the past twenty years at about half an hour in both directions.[69] Washington, like most areas, does not inflict the forty-five minutes to an hour-plus commute on its work force as is true of the New York area. In a rational world, teleworking would be concentrated among the worst commuter sufferers--the hour-plus workers found in every area. More than that, if work travel can be drastically curtailed, the current pattern could be turned on its head: instead of choosing a compromise location to balance work and residence, the choice of community would take precedence for workers who either didn't need to travel at all to headquarters offices or whose journey to distant workplaces was infrequent. Hence the predictions of an amenity-driven competition for upscale people with a wide range of geographic choices.

We can add one caveat to the prediction that new communities in exurbia will be successfully marketed to a sizable segment of new home buyers. Based on past history, the first settlers and retirees are happy in their choice. Teenagers and young, single adults tend to be bored, restless and sometimes downright hostile.

The Perils of Nostalgia or Why People Leave Small Towns

Not everyone is happy with the notion of living in the kind of small town life that housed much of the nation's population through the 1920s and 1930s. Sinclair Lewis and Sherwood Anderson are not the only observers who found the small town stifling and bigoted, especially for older teenagers, young adults, intellectuals and in fact anyone who did not fit the prevailing mode.[70] Clarence Perry, a Guru of neighborhood planning was cognizant of the shortcomings of this communitarian approach. He wrote in the 1920s that

> ...A young New York businesswoman was seeking a home for herself and mother. Upon being told about the attractiveness of the social life in a certain suburban development, she exclaimed, 'Oh, but I don't want to go where I have to know people!' What she really objected to was living in the kind of community where her daily comings and goings would be subject to the scrutiny and comment of neighbors. She recoiled from the tyranny of small-town gossip.

Thousands of like-minded young people have sought with eagerness the freedom and anonymity of rooming-house life in a large city.[71]

But Perry was also aware that time, aging and children lead to changes in preferred life styles. These same bachelor men and women experience a change of heart when they marry. Especially after the children have begun to arrive, do they again long for the detached house and yard, and the social benefits of a congenial neighborhood.[72]

Robert Lynd, the co-author of *Middletown* and *Middletown in Transition* which examined Muncie, Indiana in the mid-1920s and again the mid 1930s echoed this sentiment. In a reprise of the medieval German slogan *Stadt Luft Macht Frei* (City Air Makes You Free). Lynd observed that

> Many of those who migrate to our larger cities pride themselves on the fact that "Now, thank God, I don't have to know my neighbors, go to Rotary, belong to a church, or participate in an annual Community Chest drive!" And the big city does little to disabuse them of this attitude.[73]

The objective of the neo-traditionalist of the 1980s and 1990s and before them the newtown planners in Britain and the U.S., is the recreation of just this kind of neighborly community. Lewis Atherton describes in vivid terms the closeness of the kind of idyllic small town life they hope to recapture.

> Village people rose early in the morning and set a pace which saw them through a long working day without exhausting their energies. A leisurely tempo with slack periods gave time to enjoy others and to engage in talk, the most pervading of all social activities. Women deserted their canning, washing, and housecleaning to gossip over the back fence or to rock in another's home while they discussed departures from routine patterns of neighborhood and town life. Marriage, birth, accidents, and death were common topics of conversation. Reports on those ill circulated each morning, and rumors of moral derelictions were passed from home to home. Retired farmers, down town for the morning mail, discussed crops and weather, which had shaped their daily activities for so many years, and then

deaths and marriages. These were fitted into family and community relationships. Ancestral backgrounds, family connections, property holdings, and highlights of the career of any recently deceased member of the community were recalled and placed in their proper niche in the oral history of the village, thus giving a sense of continuity. [74]

What's missing in this happy picture? Closeness and intense interaction with neighbors was great for many families but stifling for non-conformists and confining for the teens. It's no accident that by their teens many youngsters developed an itch for bright city lights where excitement awaits, and the censorious and gossipers are far, far away.

There is a long list of graphic fictional portrayals of young people who found small towns stifling, neighborly friendliness akin to prying and moralizing, pervasive demands for conformity excessive: to the political radical, the person uncomfortable with settled heterosexual life patterns, and most of all, restless young people in search of freedom the big city offered freedom to choose without nosy, neighbors.

In past years youngsters on Nantucket, a tourist paradise off Cape Cod, complained to the author that this idyllic island was an uncomfortable place for teenagers. They were counting the days when they could desert "this damn prison" for the excitement of Boston.

Do telecommunications make a difference for these unhappy kids and their peers in small exurban towns? Certainly, cable and video players are helpful in reducing the sense of isolation but active teenagers want to dance with strangers: even video-conferencing and bodysuits won't make that possible.

The Search for Community

Some types of communities have been famous for their rigidly imposed morality. For example, there is a long history of successes and failures in founding new communities based on a shared vision. Many of the early settlements in the U.S. in the colonial era were based on religion. Persecuted sects came to Massachusetts and elsewhere in the New World to build a utopia in the wilderness. In their turn the "theocracy," Thomas Wertenbaker's term for the politico-religious rulers of Massachusetts who had long struggled against the established

Anglican church in England created a reformed religious structure which they intended to protect from innovation by the force of civil law.

As Wertenbaker says, "The Puritan community thought that heretics should only have the opportunity to leave...no man was to vote who was not a member of the church."[75]

Incursions by religious nonconformists like Quaker missionaries were fiercely resisted. In 1650 the Bay Colony's moral majority decreed that severely whipping Quakers was not enough:

> ...the offender's tongue was to be bored with a hot iron, his ears cut off, he was to be banished, and if he returned, to be executed. Rhode Island, under the leadership of Roger Williams owed much of its early expansion to its avowed toleration of Quakers and other dissenters from the prescribed theology.[76]

Religious or utopian communities with a philosophical flavor were established by the dozens in 19th century America. It was the time of New Harmony, Brook Farm and Oneida--a period when utopian visions flourished in scores of religious and economic experiments in communal living. In part these communities represented a rejection of the greed and materialism and, indirectly, of the city.[77]

(For some reason, the most successful religious-based 19th century community is usually omitted from the list, i.e., Salt Lake City). Of course religious-based settlements need not be urban: the Amish offer living proof of the survivability of closely-knit rural clusters.

If we set aside the utopian impulse, settlement based on congeniality of outlook and social affinity has been a hallmark of suburban and exurban development. Over time, people tend to seek out, then locate in communities where there is a narrow range of class differences. And unlike religious-based communities which are frequently bedeviled by schisms and charges of heretical incursions, middle- and upper- income suburbs may have an occasional internal battle over taxes, schools or other civic issues but they tend to be tranquil places to live, tolerant of people who buy in and maintain their property. In contrast, new settlers in exurbs dominated by moderate income, insular people with a parochial, sometimes intolerant, outlook may find the welcome less benign.

This does not mean that the utopian ideal has vanished. Attempts at do-it-yourself Edens flourished during the 1960s when some of the young and middle aged created small settlements based in part on the same revolt against the prevailing materialism that had animated founders of utopian communities over a century earlier. Most foundered on familiar rocks and shoals: personal rivalries, lack of income, defections by the disillusioned or restless. It is important to recognize that the beat goes on. In the 1990s there are dozens and dozens of attempts to secede from the mainstream and create small, affinity communities. The generic term is "cohousing." For the most part, these are very small--six to perhaps twenty to thirty families. There are thirty co-housing projects in operation

> ...and another 200 or so in various stages of construction or development...All together, there are more than 1000 intentional communities in America, 180 of them are well established and have existed for a decade or more. The basic concept is simple: In most projects, residents own their own housing in much the same way as condominium owners, and share ownership of common areas.[78]

The objective, based on successful Scandinavian models, is to provide a sense of belonging without forcing residents to sacrifice privacy or individuality. Projects are designed to offer an extended sense of home and family and recreate support systems that have eroded in America over the past fifty years.[79]

One major difference between the cohousing communities and the earlier utopias is the need for a substantial stake to buy in. Take the specs for a new exurban community outside Boston on eleven acres: There are ten families building homes and room for fourteen more: The cost for the new homeowners ranges between $130,000 and $200,000. The mantra for success in the 1990s is come with ideals, but and bring along a minimum annual income of $40,000 to $50,000.

What this discussion suggests is that the era of self-help is far from over. Those dissatisfied with the perceived ills of current life need only find kindred spirits and build their Eden without waiting on government or private enterprise to do so. The fact that it takes a modest income to secure this kind of independence is not surprising. Shoestring operations based on hope, welfare, support from relatives or fantasy enterprises have had a short shelf life. The moral is, reliable income first, select an affinity group second. In short, for those who want to tread this road, the possibilities are there.

Turnaround Exurbs

The term "outer suburb or exurb" generally conjures up visions of rural areas invaded by outposts of suburbia.

This is an oversimplification. As metropolitan areas expand outward they overtake small industrial cities once peripheral but now beginning to be surrounded by the kind of new subdivisions and commercial development familiar to suburban residents. For some of these small cities, new times hold little promise. They are industrial slums analogous to troubled inner suburbs that face most of the problems and challenges of central cities without their compensating assets.

Some hard pressed outer cities are looking into the possibility of adaptive reuse of old, well-preserved factory buildings.

Preservation, the publication of the National Trust for Historic Preservation reports "the number of old New England factories being reused is legion. This is after all, where mill metamorphosis incubated in the 1970s.[80]

There are no figures to indicate how much of this reuse is taking place in suburbs compared to outer suburbs but *Preservation's* summary list of past adaptations probably points in the direction of upscale areas: conversions have focused on shops, apartments and research parks. Active real estate markets capable of providing tenants who can pay high enough rents to cover the costs of purchase and renovation are rarely found in isolated rural areas or for that matter in old industrial districts of central cities or in most in exurban areas.

Regional Guidance Systems

What has been described so far is a market-driven process of centrifugal growth that assumes that most metropolitan regions and state planning agencies will observe, occasionally prod and view with satisfaction or alarm free enterprise in full bloom. Much of the local reaction will be reactive rather than proactive with local government responding to developer initiatives mostly by encouraging growth of tax base and job expansion. It is distinctly different in its passivity to the guidance displayed in two metropolitan areas in North America. (When the term "North American" is used it means there's a Canadian in the house.)

The two outstanding areas are Portland, Oregon and Vancouver, British Columbia. The Portland, Oregon region is known for its statewide land use planning efforts that limit urban sprawl, its thriving downtown, and its use of mass transit to focus urban development. Unlike virtually all U.S. areas, Portland has established effective urban growth boundaries to limit the spread of exurban development.

Moreover, downtown Portland has been revitalized by integrating transit, parking, and development policies. The key to metropolitan Portland's regional system of growth guidance is effective state planning. Oregon's planning system has been credited with, among other things:
- preserving Willamette Valley farmland and slowing urban sprawl;
- providing predictability to land owners and developers;
- equitably distributing low and moderate income housing;
- making allies of the state land use and transportation agencies.[81]

One of the original statewide goals required that every city and the Portland metropolitan area have an urban growth boundary identifying where urban activities would be located in the next twenty years. The Metro Council adopted the Portland metropolitan area urban growth boundary in 1979 and has made only minor revisions in the boundary since then.[82]

There is no other U.S. model that offers as successful a history of strong regional plans that effectively counter sprawl, but British Columbia has Vancouver to serve as a model for directing and channeling suburban and exurban growth. Once again, the key is mass transit.

> Advanced light rail and passenger ferry service has been the catalyst for the development of underutilized properties in four designated regional centers. The centers now have transit-friendly office, retail, and residential development. This metamorphosis was aided by supportive zoning, government agency tenants in new developments, and a number of other policies of the provincial, regional and local governments.[83]

This integration of transit and land use is the product of a long history of regional comprehensive planning. Regional planning in the greater Vancouver Metropolitan Area can be traced back to as early as the 1920s and 1930s. This long history of regional planning has led to a vision of a system of Regional Town Centres or mini-downtowns, linked by vast and efficient public transit system, as opposed to uncontrolled suburban growth. The regional growth vision was crystallized in the historic Livable Region Plan in the early 1970s.[84] Since that time, Vancouver has managed to live up to this prescient goal.[85]

The prospect for adopting similar systems of strong regional guidance are dim in most U.S. areas. Powerful resistance has been generated in opposition to much less stringent growth management plans in New Jersey, Maryland, Florida and other states. Developers, farmers, triaged localities and most conservatives see this regional land-use transportation approach as an affront to free markets and the potential losers have a sympathetic ear in state legislatures.

The feebleness of government agencies in shaping the form of America's regions is the outgrowth of the nation's planning tradition which restricts government's role to providing the key roads and sewer lines and otherwise taking a reactive posture to developer proposals. As Peter Hall reminds us:

> ...the real core of the American system of land-use control is not planning, but zoning. But a formally separated zoning commission for each local authority area, need take no account of the plan (if any) and it is essentially a limited and negative system of control over changes in land use. By definition, zoning is a device for segregating different types of land use, usually on a rather coarse-grained basis. What zoning cannot do in practice is to stop a potential developer from developing altogether; he must be left with some profitable development of the land.[86]

Why are there chaste greenbelts and no plague of billboards and strip developments around British cities? Why are Americans so struck with the esthetic differences? It's because zoning is not planning. It is not sufficiently proactive.

> Zoning, in a fairly rough and ready way, has achieved some of the same objectives in practice as land-use planning in Britain; it has segregated land uses thought to be incompatible, such as factory industry and homes. But by definition, it cannot protect open countryside against development; that could usually be assured only public purchase as a national or State park or similar facility. In

practice, zoning is more subject to abuse than land-use control in Britain; notoriously if the landowner or prospective developer is persistent enough, he can usually get the change he wants.[87]

It is clear that regional land use plans depend on state government action because in the U.S. with few exceptions, regional planning agencies are research and advisory bodies with little implementation power. A number of states have taken action to combat suburban sprawl and channel development to areas already served by sewer and water systems and other infrastructure. Some like Maryland, have made a promising start in this direction and want to go much further.

Maryland and other states have been moderately successful in taking belated action to protect environmentally sensitive acreage and assuring public access to waterfronts. Other states attempt to manage growth by linking development to the construction of supporting facilities and services, particularly road capacity and by encouraging higher densities in cluster concepts. A number of states have adopted this type of action to preserve prime agricultural land, to protect historic sites like battlefields and more rarely to protect and enhance "viewscapes." Generally speaking, the sum total of these efforts has been a modest amount of restraint on centrifugal growth. There is a little less sprawl than there might otherwise have been but the basic pattern of outward expansion remains unchanged.

It is possible that the sheer costs entailed in responding to endless sprawl, may be turning the tide. In Maryland, Governor Glendenning's "Smart Growth" program which seemed doomed to defeat in the early part of the 1997 legislative session, appeared certain for adoption.

Without passage of the bill, a continuation for twenty years of current development patterns was predicted to transform 500,000 acres of open space. To control sprawl, the bill would withhold or limit state tax dollars to its counties for schools, roads, sewers and other projects outside targeted growth areas.[88] Rural areas would have to average 3.5 housing units per acre as a means of countering further low-density development. In short, development would continue but there would be strong incentives for clustered housing and infill in areas already served by public facilities.

Where Will We Grow the Food?

The conversion of prime farmland to urban house lots has caused considerable concern over the past two generations on the assumption that in time, we won't be able to feed ourselves with what's left, let alone provide a major export industry and sizable contribution to world needs. Los Angeles County, in California's most productive farmland is now part of megapolis and the Central Valley is well on its way.[89]

How fast is farm land disappearing? Very fast. Residential and business development was largely responsible for significant loss of farms from 1986-1996. For example, New York lost 7,000 out of 43,000 farms in the same period and Connecticut 300 of 4,100 farms.[90]

Three observations are in order: First, as was noted above, the alarm bells have been sounding for over half a century and there has been no production crisis and quite possibly there may be none if yields continue to go up in an age of technological innovation. Second, planning and zoning incentives can reduce the demand for prime farmland by mandating higher urban densities; conversion would thereby be slowed.

The third point is that urbanization need not result in a substantial loss in agricultural production if city farming can be made more popular. Cities in Asia and Africa grow much of their own food, garden plots are a major source of food in Russia and Britain and increasingly in the United States. In short conversion need not be the path to food rationing or astronomical prices for farm produce.

The Ultimate Relocation

In the past four or five years, there have been numerous stories detailing the idyllic lives of people who have made the big jump. These come in two main varieties: "lone eagles" and "colonialists."

The title of lone eagle has been bestowed on yachtsmen telecommuters, operating in the Caribbean, on mountaineers, and sport fishermen operating solo in remote retreats but in instant, constant touch with offices, markets and subordinates. Naturally, most of these people are bosses, consultants and independent executives giving orders and advice, preparing and analyzing reports, parsing data and visual materials. The prerequisites are computer, printer, modem, satellite, phone and FAX. The temperament is, as the sobriquet implies, a lone operator who has little need for daily or even weekly face-to-face contact with a group of peers. Not many fall into this category. Much publicity, few takers. Perhaps a fraction of the labor force can break away in this fashion and an even smaller percentage has the discipline and inclination to follow this path. Of course, when two way visual contact becomes widely available (followed by virtual reality a few years later), the fraction may be considerably larger.

Rich Tax Refugees

In the mid-1990s there was a minor controversy involving a dozen aggrieved multimillionaires emigrating to tax havens to enjoy American fortunes abroad without the onerous attentions of the IRS.[91] More popular among the not-so-rich are "offshore jurisdictions" that offer discreet banking, favorable tax laws and, for the truly disillusioned tax-dodger, opportunities to acquire outright citizenship. A number of nations are in the business of luring wealthy American tax refugees to havens in Europe, Asia and Latin America. Close to home are Caribbean independent to quasi-independent countries like Bermuda, the Bahamas, the Cayman Islands and Turks and Caicos. the British Virgin Islands, Anguilla and Curacao, and St. Maarten are other oases. In Europe, there are Gibraltar and the Channel Islands; off New Zealand, the Cook Islands as well as Ireland.

Clearly this kind of legal tax avoidance has been going on for a long time but as with other types of commuting, technology makes it easier by permitting distant expatriates to work, consult, invest and stay in touch with relatives and U.S. entertainment and news media. In short, the existence of instant telecommunications serves to intensify and accelerate existing trends.

Obviously, the same strictures apply to these offshore destinations as to the similar distant honey pots in the U.S.. They tend to offer third-world or at best, second-world services and unless one is truly wealthy, frequent face-to-face contact with kith and kin is prohibitively expensive.

A favorite example of the freedom to choose is the lone eagle sailing in the Caribbean, operating a business while contemplating a sunset over St. Thomas. Lucrative labor in a playtime setting is the ultimate elysium--the outer limits of telecommuting. A "virtual manager and yachtsman" relies on the telephone rather than e-mail. Business correspondence: twenty letters a day, fifty phone calls a day, ten times a day communication with a land-bound, front office secretary.

An abandoned sugar cane field in Hawaii is the site of a software quality text center for Verifone. Hawaii offers a time zone advantage, "a graveyard shift without upset body clocks" to ship test results to the U.S. East Coast. Are employees frustrated by life in a remote location 25 miles from the nearest small town: 3,000 miles from San Francisco? No. "They get their big-city fix at trade shows."

We can offer three generalizations about these long distance teleworkers:

1. *There are very few of them.* "Lone Eagles" is an accurate title. They receive considerable publicity to compensate for their rarity, partly because some of them are smug, tanned braggarts only too happy to share with the world the story of how their extraordinary talents enabled them to beat the system, directing businesses, making sales, offering consulting advice from their yacht in the Caribbean, their mountain aerie next to the ski slope or their villa in the south of France. Vacationing reporters are always primed for this kind of "gee whiz" as in the envious, ... of the rich and famous. The tools: fax, phone, modem, computer, Fedex are no different from the complement in the suburbs and exurbs.

2. *They don't bring the kids.* Lone eagles seem to be divorced and separated from family groups. If they do have children they are either grown or living with a wife or ex-wife in a conventional suburb. This is appropriate: yacht basins or ski chalets are not as a rule, located near a good public or private school system.

3. *They enjoy autonomy--but few local peers.*

Naturally, the long range operators are mostly affluent. They have the experience and reputation to retain existing clients and to attract new ones with minimal, periodic face-to-face contact. And there is very little company other than the service personnel and staff of small local businesses. Visitors may find themselves pressed into service as instant companions. CNN and other cable compensate somewhat for the skimpy local media and the stunted local library.

Conclusion

Overall the centrifugal movement of people and businesses is likely to pick up steam with the help of new telecommunications technology.

The big winners:
- Exurbs located within two or three hours of large urban centers;,
- college towns--the real McCoy; and
- recreation areas, particularly those that enjoy a balmy climate.

In contrast, lesser winners such as far out areas like Caribbean tax havens or distant paradises in the Indian Ocean, will attract a handful of semi-eccentrics.

It seems clear that these home offices, and the small office buildings that provide space for a handful of teleworkers represent targets for economic development in coming decades. When this realization strikes home, the amenity wars will pit communities and counties in rivalry to attract upscale residential growth. The new affluents not only generate sales volume for existing local businesses but attract new health food, bookstores, coffee houses and home furnishings establishments. But most important, some of the work-at-home and telework locally newcomers will found new brain-oriented firms that provide jobs and tax base with little of the need for infrastructure or the effluents associated with traditional industry.

I am in agreement with OTA's prediction that the prime beneficiary of the new technology will be a growing number of outer suburbs. The latter offer cheaper housing and perhaps more greenery for the price and many will offer acceptable school quality. OTA cites planning research, indicating that most people (80 to 90 percent) prefer to live in low-density, single family housing 30 miles or less from a major city. Urban amenities are available in the distant suburban ring through better septic systems, satellite dishes, and TV home shopping and ready access to regional shopping centers.

1. Robert Cervero, *Suburban Gridlock*, Center for Urban Policy Research, (New Brunswick: Rutgers University, 1986), 9.
2. Carey Goldberg, "Alarm Bells Sounding as Suburbs Gobble Up California's Richest Farmland, *The New York Times*, June 23, 1996.
3. *Bay Journal,* October 1996, 3.
4. Ibid, 4.
5. Timothy Wheeler, "Coalition to Fight Urban Sprawl, *The Baltimore Sun*, October. 24, 1996, 3
6. "The Old Picket Fence and Corner Store," *The Economist,* November. 25th--December. 1st, 1995, 28.
7. Roger R. Stough and Jean Paelinck, "Technological Driven Dynamics of Outer Metropolitan Areas: Analysis and Modeling," Paper presented at Regional Science Association, Binghamton University, April 16-27, 1996, 2.
8. OTA, op.cit., 194, 218
9. Brower, op.cit., 149.
10. J.S. Davis, A.C. Nelson, and K.J. Dueker, "The New 'Burbs'," *APA Journal*, Winter 1994, 46.
11. Ibid
12. Ibid.
13. Arthur Nelson and Thomas W. Sanchez, "Exurban and Suburban Households: A Departure from Traditional Location theory?", *Journal of Housing Research*, Vol 8, Issue 2, 1997, 264,265, 268.
14. Ibid., 251
15. Ibid., 46-47.
16. John S. Bennett, cited in James Brooke, "Colorado Tries to Keep Lid On Population Boom," *The New York Times,* November 5, 1995, 18.
17. Ibid., 13.
18. Witold Rybczinski, "Big City Amenities. Trees. High-Tech Jobs. Cappacino. Retirement Paradise. Nose Rings: The Rise of the College City, The Best New Place to Live," *The New York Times Magazine*, September 17, 1995, 58-61.
19. City of College Park, *Comprehensive Plan Preliminary Draft*, October.20, 1995, 7-2; 7-5.
20. Hubert S. Stroud, *The Promise of Paradise: Recreation and Retirement Communities in the United States Since 1950*, (Baltimore: Johns Hopkins, 1995).
21. U.S. Department of Commerce citied in *Macmillan Visual Almanac*, (Blackbirch Press, 1996), 405
22. U.S. Bureau of the Census, *Statistical Abstract of the United States, 1996*, Table 388, 2240.

2 3. Haya El Nassen, "Recruiting Retirees," *USA Today*, May 3, 1996, 3A
2 4. Ibid.
2 5. Ibid
2 6. Suzanne Sutro Rhees, "Mall Wonder, *Planning*, October 1993, 18-23
2 7. Japan National Tourist Organization, May 1, 1996 (invitation letter)
2 8. Michael D. Shear, "In Prince William Development Plans Run Into Opposition: Developers," *The Washington Post*, February. 18, 1996, B9.
2 9. Ibid., B10
3 0 .Ibid.
3 1. Timothy Egan, "Urban Sprawl Strains Western States," *The New York Times*, December. 29, 1996.
3 2. Timothy Egan, "Drawing a Hard Line Against Urban Sprawl," *The New York Times*, December. 30, 1996, 1.
3 3. Ibid
3 4. Ibid
3 5. Ibid
3 6. Tamar Lewin," In Home Schooling, A New Type of Student," *The New York Times,* November 2, 1995. D-20.
3 7. Rene Sanchez, "Embracing New Schools of Thought," *The Washington Post*, December. 5, 1995, 1.
3 8. Kate Zernike, "Class Struggles," *The Boston Globe*, June 10, 1996, 1,20.
3 9. Neil MacFarquhar, "Public, but Independent, Schools are Inspiring Hope and Hostility," *The New York Times*, December. 27, 1996, B-5.
4 0. *Plaza del Rey Newsette*, (Sunnyvale, CA) May 1996, 4.
4 1. "Farmers Cultivating Suburban Neighbors," *The New York Times*, February .2, 1997, 21.
4 2. Evan McKenzie, *Privatopia Homeowner Associations and the Rise of Residential Private Government*, (New Haven: Yale University Press,1994)
4 3. Ibid.
4 4. Roger Lewis, "Gated Communities Shun Diversity," *The Washington Post*, September. 9, 1995,E1,12.
4 5. David Boaz, "Gates of Wrath," *The Washington Post,* January 7, 1996, C.2
4 8. Alvin Toffler, *Power Shift*, (Batan, NY, 1990), 73-74.
4 7. Ibid.
4 8. Graham and Marvin,op.cit., 222-223
4 9. Kunstler, op.cit., 263
5 0. Kenneth Lelen, "Elements of Style: Builders Find Neo-Traditional Designs Have Their Limits," *The Washington Post*, June 22, 1996, F.3.

5 1. Mona Damosh, *Unvented Cities: The Creation of Landscape in Nineteenth Century New York and Boston*, (New Haven: Yale: 1996) 103-104.
5 2. John Rothschild, *Time*, December. 5, 1995, 62.
5 3. Ibid, 63.
5 4. Ibid.
5 5. Ibid.
5 6. "American Survey", *The Economist*, November 25, 1995, 27.
5 7. Ibid, 28.
5 8. Ruth Eckdish Knack, "Once Upon a Town", *Planning*, March 1996, 13.
5 9. Kenneth E. Harvey, "Home Buyers Want ModernTowns with Lots of Old-Fashioned Charm," *The Washington Post*, June 1, 1996, F1.
6 2. .Ibid.
6 1. John Henry, " Is Celebration Mayberry or a Stepford Village?," *Professional Builder*, September 1996, 47.
6 2. Ada Louis Huxtable, "Living with the Fake and Liking it," *The New York Times*, March. 30, 1007, Sec.2, 1.
6 3. Randy Warner, quoted in William H. Lurz; "New Communities," *Professional Builder*, Septtember 1996, 16
6 4. Jennifer Lenhart, "New Loudoun Communities: Growing Up Fast," *The Washington Post*, November 16, 1997, B1.
6 5. Ibid,B-5
6 6. Laura Zelenko, "Traffic and Pollution Create Satellite Offices," *American Demographics*,June 1992, 26.
6 7. Kenneth Lelen, "Builders Offering Homes Offices," *The Washington Post*, March 9, 1996, E-2.
6 8. Ibid
6 9. David M. Levin and Ajay Kuman, "The Rational Locator: Why Travel Times have Remained Stable", *JAPA Journal*, Summer, 1994, .323-329.
7 0. See Sinclair Lewis, *Main Street,* (New York: Signet, 1920, 1961). Also Sherwood Anderson, *Winesburg, Ohio,* (New York: Viking1966) (First ed. 1919).
7 1. Clarence Arthur Perry, *Neighborhood and Community Planning*, Vol.VII of Regional Survey of New York and its Environs (New York: Regional Plan of New York and its Environs, 1929), 25.
7 2. Ibid.
7 3. Robert S. Lynd, *Knowledge for What? The Place of Social Science in American Culture* (Princeton, NJ: Princeton University Press), 83.

7 4. Lewis Atherton, " The Small Town in the Gilded Age," John H Cary/Julius Weinberg, *The Social Fabric,* (Boston: Little Brown: 1984), 59.
7 5. Thomas J. Wertenbaker, *The First American: 1607-1690*, (Chicago: Quadrangle: , 1927, 1955) 97, 101.
7 6. Ibid, 102.
7 7. Charles N. Glaab and A. Theodore Brown, op.cit.
7 8. Skye Alexander, "Cohousing in New England," *Earthstar*, June/July, 1996, 38.
7 9. Ibid, 39
8 0. "Trial By Fire," *Preservation*, November/December 1996, 78.
8 1. Report 16, *Transit and Urban Form Transit Cooperative Research Program*, Vol.2 *Public Policy and Transit-Oriented Development*, "Portland Case Study," 71, (Washington: D.C.: National Academy Press, 1996)
8 2. Ibid, 80 (Note: The 1990 Census reported that 3.4 per cent of the Portland labor force and 3.8 per cent of suburban residents worked at home. (74).
8 3. Ibid, 84.
8 6. TCRP H-1 Public Policy and Transit Oriented Development, "Vancouver BC, Case Study," 99
8 5. (1995 Canadian Dollars [CAD]; GVRD: *Greater Vancouver Key Facts*).
8 6. .Peter Hall, *Urban and Regional Planning* (London: David and Charles,1975,) 262
8 7. Ibid.
8 8. Terry M. Neal and Todd Shields, "Md. Looks to Regulate the Promised land," *The Washington Post*, February. 23, 1997, 6
8 9. Carey Golder, "Alarm Bells Sounding as Suburbs Gobble up California's Richest Farmland," *The New York Times*, June 20, 1996.
9 0. "Farmers Cultivating Suburban Neighbors, *The New York Times,* February .2, 1997, 21.

Chapter 4

Dreams and Nightmares

The optimistic prognosticators who follow the great American tradition of seeing the upside in new inventions have greeted rapid advances in telecommunications technology as the pathway to all sorts of joy to come. Following the yellow brick road to Oz will find the decrepit, the frail and the disabled granted the gift of mobility and employability. Persons trapped at home as caregivers for sickly elders or young children will be enabled to hold part-time or full-time jobs. Everyone who can pay the small entrance fee for the requisite equipment will thereby enter a fascinating world of entertainment, information and constant contact with friends, relatives, colleagues and fellow members of interest groups. The educational impacts will be breathtaking when the gifted can be stimulated to new heights of achievement and mediocre and laggard students can be energized with new forms of experiential learning. And as, the preceding text has suggested, broadened by use and familiarity with telecommunications, many people will be offered a wider range of residential locations and job opportunities. Places widely acknowledged as desirable places to live--select suburbs, many college towns, a number of exurbs and recreation areas--are likely to find themselves in the direct path of telcommunication-assisted relocators with the concomitant pressures on real estate values, roads and other facilities and services required to accommodate the new population and new economic growth. At the same time we can anticipate more bad news for many disfavored and inner suburbs as more middle income people leave (although gateway cities may compensate for their losses by attracting niche markets and foreign immigrants).

At a mundane but critical level, telecommunications may serve as a substitute for much highway travel. This may have substantial, beneficial impacts on air pollution and highway congestion.

Less Congestion

How much of a reduction in daily commuter travel can we anticipate as a result of teleworking? Mokhtarian offers preliminary estimates based on California's experience. The assumption is that 16 per cent of the work force can now telecommute. For a variety of reasons including change in employer attitudes from hesitation and resistance to mandates, the proportion of the work force telecommuting on any given day will rise from a minuscule 1.5 per cent in the early 1990s to a still tiny 2.7 per cent at some point in the future.[1] What this estimate omits is the large number of part time telecommuters who travel off peak, thereby contributing substantially to relieving congestion from the highest volume.

Mokhtarian does point out that time saved by telecommuting may be used for other travel, that telecommuters may move farther away from work, thus necessitating longer travel times and distances and that freeing up some highway capacity by telecommuters may cause other drivers to fill the vacant space.[2] What this cautious appraisal fails to take into account is the snowball effect. For example, a growth of new suburban close-to-home telecenters many of them requiring short commutes on back roads, will greatly diminish peak hour travel on congested major highways. This powerful friction avoidance on the part of employees as well as an alteration in employer attitudes will improve the bottom line: Headquarters office space and parking is expensive. Telecommuting even at the cost of $5,000 to $10,000 or so in initial start up costs per employee is much cheaper. Mokhtarian's telecommuting percentages in the early 1990s were low because 51 per cent in her sample reported "manager unwillingness" as a key external constraint.[3] A changeover from reluctance to managerial eagerness would produce major results. This changeover is highlighted by *Business Week:*

Work anywhere, anytime is the new paradigm...it amounts to a massive disaggregation of work spinning the walls and confines of the traditional office.[4]

The magazine estimated in 1996 that 83 per cent of U.S. companies were now embracing alternative office strategies. The American office is evolving rapidly in two directions. The first is reorganizing workspace for those employees who must still work in offices and the second is "shoving everyone else out the door."[5]

Thanks to phone FAX and computer (including electronic banking and bill paying) millions of highway trips are not taken. Communications is already a substitute for an enormous amount of automobile travel. We can only speculate on the extent to which creeping highway gridlock will stimulate more employers and more employees to take active measures to avoid the peak-hour trips consuming so much time and, in spite of cellular phones and learning tapes, are so much less productive than office time undistracted by stop-and-start road travel.

Other Joys

Edward Cornish, the futurologist, offers a number of happy possible outcomes of the new technology: more time to contact distant friends; further growth in rural and resort areas; homes so attractive, so snug and safe, so full of entertainment and education resources that people won't want to leave them.[6]

Another futurologist, Joseph Coates, foretells the rosier possibility that the number of couch potatoes will diminish when two-way interactive technologies become commonplace. The quality of life will improve via expanding cultural opportunities. He says, "When given a choice, many will choose interactive media over watching television." Moreover in the radiant future:
- Throughout the advanced nations, people will be computer literate and computer dependent.
- Worldwide, there will be countless virtual communities based on electronic linkages.

Planners will reallocate terrain and physical space to make more-- efficient use of scarce resources. In other words, cities will be redesigned and rezoned to improve efficiencies of energy in transportation, manufacturing, and housing, etc.[7]

While there can be no quarrel with the first and second statements., we can take issue with the other two. Will people be computer literate throughout the advanced nations or will there be an underclass of workers permanently unemployed who will lack basic literacy and numeracy, let alone computer skills? Also the notion of redesigning cities on the basis of presumed efficiency runs counter to the trend toward redesign by personal choice. Human factors, the pull of amenities and the decisions to avoid the ugly and the hazardous are likely to be more significant in the coming generation than efficient transport of goods and resources.

Is There a Downside, a Gray Shadow on the Rosy Glow?

Past history teaches that each new technological advance creates turbulence in its wake, displaced workers with obsolete skills, outmoded farms and manufacturing plants and on occasion, the threat of mass annihilation from new methods of warfare. In this context, telecommunications engenders an anticipation of disturbing changes which are not directly linked to the technology but to the fact that this form of communication will further accelerate tendencies toward fragmentation already in full bloom. .The Unabomber is not the only individual who fears that unchecked and undirected, rapid technological change has its fearsome side.

What is it we find unsettling? We can offer three major nightmare scenarios:

- Further fraying of bonds in communities and employer-employee relations.
- Moral and psychological breakdown of the most severely stressed, unstable elements in society.
- Large scale redundancy: people unwanted, unneeded in the workplace or anywhere else.

One scenario calls for a dehydrated employment structure with a few big winners and large numbers of smaller winners, many of them edgy about their precarious hold on the upper reaches of the job ladder. And below the top ten or twenty percent there will be (already are) a substantial middle-class fearful of their and their children's future in a turbulent economy. At the bottom there may well be a fifth of the population consigned to low paying jobs, many temporary, only a paycheck or two removed from foreclosure and welfare. In contrast to Marxist prophecies, this is not a prospect of exploited hardworking proletarian wage slaves but a future of permanent redundancy, of idleness and welfare for many. West Europe in the 1990s offers a preview of seemingly intractable unemployment rates of ten per cent plus and much higher rates for youth and new entrants into the labor force.

Apprehension over the impact of the new technology exacerbates existing fears that major social elements are weakening, moral foundations crumbling, economic, community and family structures failing. Within this context and the existing ills of the society in general, new technology is seen as an additional destabilizing factor contributing to the disruption.

We can identify seven separate areas that have provided the raw material for political rhetoric, religious exhortation and scholarly study:
- The decline of traditional family structure;
- loss of community and neighborhood ties;
- weakening influence of religious and voluntary associations;
- fear that public school standards are slipping;
- fear that street crime will impinge on middle and income areas,
- possible large scale unemployment;
- domination by one major interest group--big business--with the further weakening of labor union membership.

Mass Unemployment Ahead?

Predictions and fears that the introduction of a particular form of new technology would lead to heavy, permanent unemployment date back at least to the early nineteenth century when the English Luddite broke the weaving machinery that they feared as a short cut to the industrial garbage heap Similarly, each new technology that promised to wipe out jobs was seen as a herald of imminent disaster. There was for example an "automation" technology-work-reduction scare in the Kennedy Administration when some prophets of doom were predicting that any job calling for an IQ of less than 80/or 90 could easily be replaced by a machine.

A new fear is the overseas competition in which jobs can easily migrate to Mexico or Asia either by outsourcing or by using communication technology to perform white collar work for U.S. firms in low-wage overseas enclaves.

The principal difference between the current situation and previous generations is the shift up the education scale. Many of the jobs that are being created in place of those being lost in the process that Joseph Schumpeter aptly called "creative destruction," require higher levels of education and training than was the case when armies of high school dropouts could find a good paying berths on auto assembly lines or in steel mills. We have always had a have and have not economy. The difference is the expectation that people with a poor educational background will have a harder time breaking out of the ranks of the have-nots.

Exploitation, No? Redundancy, Yes?

For generations the plight of wage slaves has provided ammunition for advocates for the downtrodden. The equation was simple: workers produced surplus value most of which was captured by the owners and managers of capital and the laborer was given a sufficiency for his care and feeding. Were he foolish enough to complain he could easily be replaced by willing hands drawn from the reserve army of the unemployed or by imported, desperately poor alien labor. To ease his misery, there were unions and government who provided the working-class with countervailing power to improve pay, job security and safety.

From time to time the possibility of replacing labor with technological devices arises, the new machinery is introduced in farming, in mining, in manufacturers--and before long there are more jobs than ever and wages rise. Is this happy ending in store for the latest go around with telecommunications?

In the first place, one nightmare of shrinking employment seems overdrawn: we have been creating jobs in huge numbers. The prime cause for disquiet seems to lie in other areas: diminished job security and the plight of the poorly educated underclass.

The decline of union membership, the growing reluctance of government to intervene in the workplace, and the diminishing prospects for long-term employment with a single employer are all facts of life in the 1990s. In this emerging world of free agents, (people unprotected except by their marketable skills in rapidly changing, highly competitive environments), we can foresee an acceleration into a two-tier economy that has long existed. In past generations the automobile assembly line worker with a high school diploma could earn as much as many college graduates. Since the early 1970s, the earnings gap has been widening between the college graduate and the people without this key credential.

What we can look forward to is more polarization: The anxious swimmers in the mainstream earning sizable sums, much or most of the remainder happy to keep K-Mart level jobs serving the affluent at little more than minimum wage. To make, it they will have to pool two incomes, moonlight at a second job and pray for good health and good luck. If past experience no longer holds true, increasingly many will become redundant, unneeded, replaced in part by machinery, and by imported labor. They will not be exploited; exploited workers have a place in the economy, lowly as it may be. They will be like the long-term unemployed in Britain in the 1920s when Winston Churchill suggested that 10 million Englishmen emigrate to other Dominions since there would never be enough jobs for them in the U.K. Similarly Middletown's business class in the depths of the 1930s would have been happy to see surplus working-class people go someplace else.[8]

The strained race relations in the late 1960s culminating in race riots in 1967 and 1968 generated considerable speculation about the future place of black employment in American society. Some of the fictional forecasts called for total extermination.[9] Other academic predictions foresaw a future of grudging welfare and a few miserable jobs for a miserable underclass, mostly black.[10]

There is no doubt that blacks are over-represented among the unwanted, the long-term welfare clients, the poorly educated marginal workers scrambling for minimum pay. But the great divide increasingly is between mainstream competitors and the marginal workers with at least one academic degree as a basic certificate of employability.

One discredited method of reducing the numbers of the feckless poor, particularly blacks, was the "Arkansas appendectomy"; sterilization operations performed on welfare mothers and persons designated as retarded. Less forceful proposals like compulsory implantation of contraceptives in welfare mothers have met with resistance from black leaders who see this as a form of genocide and from Catholic Right-to-Life advocates who call for prayer, abstinence, rhythm systems and adoption. In recent years, there is a growing conservative consensus that one way to reduce the number of the underclass is to remove financial incentives for adding out-of-wedlock children to the welfare rolls. Underlying this form of social engineering is the traditional eugenicists' fear that the intelligent and affluent are barely reproducing while the poor and unintelligent are breeding like rabbits.

The potential job devouring impact of the new telecommunication technology is described vividly by a science fiction writer, Bruce Sterling:

> It's been said that the key impact of the computer networking revolution is to collapse the costs of distribution and remove the middlemen. Federal Express, which tracks their shipments via computer and microwave links, collapses the cost of distribution and direct marketing. Electronic credit eliminates financial middlemen. Corporate e-mail collapses middle management. Unfortunately almost all of us are middlemen in something, or middle management in something. You end up with hollowed-out corporations, a collapse in job stability. Companies begin reengineering themselves with frantic speed, flinging career employees aside like so much human shrapnel. As a vision of the mid-twenty-first century, I offer you the vision of a man freezing to death under a bridge with a last decade's state of the art laptop showing the entire Library of Congress on atomic-density CD-ROM storage. Bums with a virtual world in a paper bag...[11]

Sterling's prediction for the 1990s and beyond is whether the chronic insecurity plaguing the working-class will really work its way up the food chain. In future years will a college degree offer any assurance of job security? Lester Thurow doesn't think so:

> The middle class is scared and it should be. The supports for its economic security are being kicked out from under it. ...after World War II, white collar workers and managers came to expect lifetime employment with rising wages, assuming their firms remained profitable.[12]

Supposing income inequalities grow, the state-financed safety net is weakened, unions continue to lose power and membership and social investments in education, infrastructure and research and development declines, what happens in another generation?

The spread of telecommunications technology in conjunction with reengineering, rightsizing and other "lean and mean" efforts seems to pose new challenges to previously unthreatened sectors of the labor force. If the predictions have any validity, this time it is not blue-collars under the gun but white-collar workers as well.

Up to the late 1980s employment growth in offices and services more than made up for stagnation in manufacturing employment and major losses in farming and other industries. By the mid-1990s white-collar job expansion slowed, victim of mergers and acquisitions and a relentless urge to cut costs and improve earnings. ATM's decimated the ranks of bank tellers, CADS shrunk the need for draftsmen, and new communication devices diminished the army of telephone operators--and on and on.

As Jeremy Rifkin sees it:

> ...automated technologies have been reducing the need for human labor in every manufacturing category. Now, however, the service sector is also beginning to automate: In the banking, insurance and wholesale and retail sectors, companies are eliminating layer after layer of management and infrastructure, replacing the traditional corporate pyramid and mass white-collar work forces with small, highly skilled professional work teams, using state-of-the-art software and telecommunications technologies. Even those companies that continue to use large numbers of white-collar workers have changed the conditions of employment, transferring workers from permanent jobs to "just in time" employment, including leased, temporary and contingent work, in an effort to reduce wage and benefit packages, cut labor costs and increase profit margins.[13]

So far Rifkin's dire predictions of job losses have been wide of the mark. The Clinton administration was able to boast of a 10 million employee growth in jobs in four years. The real questions concern the situation in ten or twenty years. In the midst of the furious cost cutting of the 1990s there is a new wrinkle in long range telecommunications. For about a decade U.S. firms have tapped into clerical labor pools in Barbados, Ireland and Singapore, English-speaking, disciplined and cheaper than the U.S. counterparts. Wall Street has posted offshore back office outposts, and insurance companies have shipped low-level data processing activities to places where labor costs are a fraction of the U.S. level.

Two facts must be taken into account in putting this into context. The first is Secretary of Labor Reich's contention that many high tech jobs will remain in the U.S. because electronic mail and facsimile machines are no substitute for working side by side with customers and colleagues. Despite the occasional bursts of publicity, the number of offshore engineers remains very small. In the long run the numbers may grow. If there is a critical mass of talent in Bangalore, new ideas will be generated from the source along with modest amounts of new jobs. We are not talking about lone-eagles on Caribbean yachts, but the growth of substantial new high technology centers. Moreover, India is not alone in talent pools. Russia with extraordinary numbers of underpaid first rate mathematicians and scientists is a huge source of very low-cost quality labor. In the mid-1990s high quality Russian specialists were on the market at half the monthly rate in India, perhaps a fifth to a tenth of U.S. levels.

In the 1980s newspaper stories dwelt at length on the plight of the displaced rust belt worker, down from $35,000 a year with overtime in steelmaking and auto assembly plants, reduced to food stamps, grocery bagging or clerking at convenience stores. The 1990s have had their human interest stories move farther up the line. Professionals and middle management in their forties and fifties victims of mergers, reengineering and plain old cutbacks have been laid off much like their blue-collar cousins. Tragic stories of futile job searches, ineffective attempts to hide the truth from colleagues, friends and neighbors, bills piling up, debts mounting and at last, drastic reductions in status and pay for jobs well-down the career ladder.

The fear for the future is that telecommunications skills or more precisely the lack of such will add to the woes of the hard pressed, middle-aged work force. Unlike their younger children, unlike their their job competition, this generation has not grown up with the technology, absorbing changes like an experienced skier navigating moguls but an older generation often finds the new technology intimidating.

Bullish on America?

Gloom in the United States is an ephemeral phenomenon. It comes in dark waves of anxiety during recessions and departs in the radiant sunshine of the next bull market. Indeed, past forecasts of doom have proved dead wrong. In the mid-1960s, for example, Will Ferry was filled with foreboding:

> Our growing inability to absorb available workers into the economy will give the question of 'just distribution' a point it never before had. These technologically displaced people will comprise a new class, which I shall hereafter refer to as the 'liberated margin' permanently liberated...from traditional toil, not cause they want it so, but because the imperatives of efficiency have sent them to the sideline. ...Many questions arise. How will they live? Who will provide for them? How close to reality is this vision of the automated future? How can the seemingly inevitable be avoided?[14]

In 1962, John Dunlop wrote that President Kennedy at a press conference stated that:

> ...the major domestic challenge of the sixties is to maintain full employment at a time when automation is replacing men...James Reston writes that machines are replacing everything in this country except maybe pretty girls and President Kennedy is worried about that.[15]

Other scholars writing for the Dunlop study were much more complacent. For example, W. Allen Wallis concluded that:

> ...building automation equipment though it may reduce the total number of man hours of work required over a long period has the effect of raising the employment in time... In short the problem of general unemployment from automation is a nonexistent will-o-the-wisp problem.[16]

In the mid-1960s, Edward Kalachek was also more sanguine and more prescient than the doomsayers. He concluded that "there is no evidence that automation represents a new or serious threat to full employment." However, Kalachek also erred in one respect. He maintained that it is likely that "a set of policies and a general economic flexibility have been achieved which are capable of maintaining a high degree of personal economic security in a time of rapid and uneven technological progress."[17] A generation later as the nation nears the anxious 2000s, few would be that optimistic.

It is clear that increases in production, regaining the nation's competitive position and substantial gains in profits can all be achieved without significantly expanding employment levels. For this reason investment in new technologies could result in "jobless growth" in many industries, similar to the employment losses occurring with production gains like those that have long been characteristic of mining and agriculture. Furthermore, the finance and service sectors, the sources of huge employment growth in the 1950-1980 period, proved to be vulnerable to new automated techniques in the 1990s. The word processing and computer-microchip revolution has already had a depressant effect on the demand for white collar labor. In short, after years of crying wolf every time new machinery and work processes are introduced, there may be genuine reason for apprehension in the 2000s. Gerald O'Neill summarizes the job arithmetic simply:

> At present it has turned out to be practical to take an industrial operation with a certain number of human workers and automate it so that the work force is reduced to a tenth the number or so that production goes up tenfold for the same number of employees...to go farther than that gets very expensive...the trainable welding machine...costs about as much as a year of employee time.[18]

In the midst of each business recession, Luddite talk revives, and there are a dark intimations that it is the busy machines that are keeping jobless people idle. Movies on this subject include Rene Clair's *A Nous La Liberte* (1931), which ends on a happy note as workers fish and frolic while a fully automated assembly line producing radios earns their paychecks for them. Most of the depictions of the impact of automation are less happy--the machinery earnings are definitely not allocated to joyful workers. In fact, as Charles Chaplin's *Modern Times* (1935) suggested, not only are there very few beneficiaries of the system, but the remaining work is deadeningly repetitious, viciously exploitative, or both.

Forecasts of a workplace dominated by error-free, nonunionized machinery which would replace virtually all labor requiring only minimum intelligence, not only on assembly lines but in the fields, in the mines, and in building, cleaning and maintenance are based in part on the fact that in agriculture and mining this is already a reality. The work force employed in these two industries has declined by about 85 percent in the past seventy years. But the prophets of doom have been repeatedly confounded by the rise of entire new industries--computers, television, air conditioning, aviation--and an enormous growth in white-collar services and research activities. It is this growth that is called into question, partly as a byproduct of telecommunications technology. Certainly we can take comfort from past history: In the 30 years between the automation scare of the early 1960s and the mid 1990s no less than 50 million jobs were added in the U.S. All the frightened, persuasive Cassandras were dead wrong.

Does teleworking have any specific, identifiable impact in this era of job insecurity? Is it harder to fire someone if you barely know him as a genuine human being instead of a presence on a monitor? The evidence points in the opposite direction. Enthusiastic downsizing has taken place with barely a ripple from telecommuter workers. True, monitoring sales output or other indices of output is easier on the screen but this represents a minor addition to the downsizers' arsenal. Hatchet men don't need much ammunition or incentive.

The real questions concern the long run, perhaps ten to twenty years in the future. They involve the progressive dehumanization of much of the workplace with people reduced to ciphers in virtual communities devoid of meaningful contact. We can foresee a world of home shopping networks in which consumers purchase a few moments of personalized chatter from cheery order-takers. Contact will remain with nurses and waiters, barber-stylists, and school teachers but even these may be progressively shortened. When sales clerks become an endangered species we can pose simple questions: Where are the jobs coming from? Where does redundancy end?

By the end of the 1990s there were enough warning signs to make analysts nervous about future job openings:

> If data-entry jobs actually become scarce, will an entire generation of young clerks and secretaries be bypassed before it has had the opportunity to compete? What will the organization of the future be if

most or all data entry and back-office administration is so fragmented that it resides amidst the lowest bidder anywhere in the world? And, what is the implication for the employer if he has the opportunity to move the back office off-shore with the flick of a switch? What will be an "entry-level" job in the year 2000? What will happen to all those young people not bright enough or motivated enough to qualify for a management trainee position.[19]

While polls clearly show considerable uneasiness relating to the prospects for job security in the late-1990s and beyond, there are conservative analysts who see the concern either as unjustified, transitional but necessary, no worse than in previous decades like the 1950s or simply the product of a spoiled generation, the "baby boomers." James K. Glassman offers an eloquent version of the latter view:

> ...self-centered baby boomers themselves occupy lofty positions in the press and government. The Me Generation has become the (Woe is) Me generation... Never mind, the facts or the context... so stop whining, baby boomers. You're getting older, but in a world in which brains count more than brawn, you still have an opportunity to get better.[20]

A Perpetual Age of Anxiety?

Looking back on U.S. history, there have been precious few Eras of Good Feelings (and far more decades ravaged by impending war, panics and depressions. Hard times have been knocking at the door of most American families for centuries. A run of bad luck, ill health, a gambler or drunk at the family helm, a downturn in the national or local economy would plunge the average family into deep trouble.

In this context, the five decades since 1945 have been something of an aberration. Two medium-sized wars, Korea and Vietnam, half a dozen mild recessions, three or four years of race rioting, and a threat of atomic doomsday that never materialized made for some disaffection, periodic attacks of nerves, but nothing to match the sheer despair and fear of the 1930s and early 1940s.

The question that concerns us is whether we are in for another decade or two of moderate edginess or else are we due for a bout of unsettling turmoil that may last for at least a decade? Will telecommunications technology have a moderate impact akin to the typewriter or calculator? Or will it be a volcano-like the automobile? The evidence seems to point to telecommunications as a contributor to a big bang, particularly if we add it to the other unsettling trends of the 1990s, the downsizing, outsourcing, global competition, the weakening of ties to community, union, church, fraternal orders and families. Telecommunications permits more of the same only faster and farther. And at the end of the road, with virtual reality a reality, we may get something no one really understands or can foresee. Possibly much of the population may be excluded from the new world by virtue of lack of education and skills, most of the remainder worried about job tenure and retreating into home cocoons and community enclaves. And perhaps a sufficiency of new jobs, but many low-wage and with few benefits.

One other prediction: hopes that one or another technology would remedy urban ills--schools, crime, negative attitudes--will recede into the distant past. The realization that technological fixes don't work on people problems is becoming part of a national consensus. As noted, new technology helps the elite to move away from people and places noted for their dysfunctional population and environment.

Some suggest that the major byproduct of the technological change will be further dehumanization of the workplace.[21] Others see the disappearance of whole occupations--farmers, miners, fishermen, bank tellers, telephone operators, draftsmen, typists, mailmen, bank tellers, bowling alley pinball setters, elevator operators and milkmen followed by fewer manufacturing workers and by some reduction in many middle managers. Based on past history, all we can say at this time is that the change is

- unstoppable;
- likely to have negative impacts on many, particularly older, workers in vulnerable occupations (along with many geographical trouble spots);
- will open up new opportunities for many, especially younger, adaptable workers and for selected areas; and
- in the long run may create more jobs than it destroys.

The examples of past ups and downs in transportation, railroads, automobiles, airplane vs horse drawn vehicles vs television and movies among countless others is comforting; history points to at least the strong possibility of growth or a scale that more than compensates for short term miseries.

The jokers in the deck are twofold:
- Technological change on a scale so vast, so rapid that it digs deep and burrows permanently into the nation's work structure; and
- a global marketplace that offers highly qualified labor at a fraction of U.S. cost

Both are present day realities. What is not clear is the dimensions of these powerhouse factors in coming decades. Based on past history, we can be cautiously optimistic.

> Always there have been winners and losers during such times of change, and the losers did not deserve their fate any more than today's middle manager does, yet in the end we've always managed to adapt to the economic forces around us. Yes, the old contract between company and employee is dead. Yes, corporate loyalty will probably cease to exist, but eventually some new ethos will replace those values.[22]

Certainly, it is true that the harsh 1930s were followed by the booming '50s and '60s and so it has been and will always be--in the view of the spokesmen for corporate America. Fortunately, history is on the side of the optimists.

One caveat: While past experience suggests that there are grounds for confidence in the future for most people and for selected areas, there are dissenters. For example, unlike high tech optimists, Graham and Marvin see trouble ahead particularly for the disadvantaged people and places.

> As high rates of structural unemployment continue in western cities-- with attendant social and political crises--these processes of change raise major concern bout the future of urban economies. How sustainable is an urban economy when it has been so fragmented, exposed to global telematics--based networks, and pitched into the global battle for investment? Is there an employment future in telematics-based services for the western cities so devastated by the continuing collapse of manufacturing employment?

> Despite the efforts of many from telecommunications and corporate worlds to suggest that the urban economic future is rosy, it is hard not to conclude that the answers to these questions--from the current perspective at least--look rather bleak. The current restructuring

processes are driven by the imperative of reducing costs, employing fewer permanent staff, replacing labour with capital, increasing automation, and thus supporting greater responsiveness, flexibility and competitiveness. Much heralded innovations such as 'back offices' and 'teleworking'--while bringing undoubted benefits to those cities and groups that gain access to new employment through them -- often only serve to redirect the jobs that remain following restructuring as they are shifted to cheaper locations or cheaper and less secure employees.[23]

Root Causes: The Vanishing Two-parent Family

Whatever the problems, the people and places who fail to share in mainstream progress and are beset by an interlocking set of problems: schools, crime, welfare, unemployment. Much of the genesis lies in single parent families, particularly poor mothers raising children on welfare or marginal incomes. Most single mothers live in poverty and their children provide most of the recruits for the criminals, the substance abusers, the school dropouts and the chronically unemployed. Cause and effect are the subjects of much debate. What we do know is that the same ills are found elsewhere in European cities for example, and there seems to be a link between rising rates of illegitimacy and the decline of civility and livability in central cities.

Nicholas Eberstadt points out that the trends are not all unfavorable. For example, the percentage of at least once-married black women in their early thirties declined from 87 per cent in 1975 to 61 per cent in 1990 But, the fact remains that out of wedlock children are likely to be poor, spend time on welfare, more likely to be involved in crime than children of two-parent families and produce out-of-wedlock children themselves.[24]

If the trends of the past thirty years continue we are apt to see in the U.S. a pattern common in parts of Western Europe: Many single, divorced or separated women raising one or two children with the help of female relatives and friends. Marriages, if they occur, last a few years and men in general seem to be regarded by self sufficient women as self-centered adolescents, prone to instability, alcoholism, unreliable as partners and useful only for recreational purposes.

One of the critical changes in U.S. and, indeed western, values is the relaxation of the social, cultural and moral stigma against out-of-wedlock births. William Galston, a former domestic policy adviser to the Clinton administration points out that

> ...poll data indicate that about 56 per cent of all Americans believe that people who generate a baby out of wedlock should not be subject to moral reproach of any sort. Among young people age 18 to 34, fully 70 per cent say: no reproof, no judgment.[25]

The relaxation of moral standards has its price: A child born before his parents finish high school, reach age twenty and get married is almost guaranteed a life of poverty.

The weakened link between parenthood and long-term marriage is as apparent or more so in Western Europe as it is in the U.S.. No one is happy with this trend. All the evidence indicates that children in single parent families are worse off than children in two parent families not only in terms of higher rates of poverty but in poorer school achievement, and in emotional and physical well-being being. But it is in the U.S. alone that out-of-wedlock births seems to be linked to severe dysfunction, particularly high rates of delinquency.

How does the U.S. stack up in out-of- wedlock births? Rates in Iceland and Sweden, Denmark, France and Britain are all higher. Where the U.S. leads is in the proportion of one parent households with missing fathers. The U.S. rate is 25 per cent, Sweden less than 20 per cent. One important difference: In Western Europe out-of-wedlock childbearing is mostly centered among mature women. In the U.S. one birth in eight is to a teenager.

> Are generous government welfare benefits partly responsible for illegitimacy as some conservatives charge? In the U.S., child benefits are low--the government spends four times as much on the elderly as on children, although old people are much better off. Lone parents are six times more likely to be poor than a married couple. Therefore, the welfare cheque, with its attendant medical benefits and food stamps, become considerably more valuable to have than a low-wage husband/father.[26]

In the midst of long-term poverty and minimal prospects for improvement, a predictable source of income, however small, is preferable to inconsistent help from a partner whose earning ability does

DILBERT reprinted by permission of United Feature Syndicate, Inc.

DOONSBURY © (7-25) G. B. Trudeau. Reprinted with permission of UNIVERSAL PRESS SYNDICATE. All rights reserved.

not match his sexual potency. What happens to children raised by welfare mothers?

For boys the absence of men can induce what sociologist Elijah Anderson calls "hypermasculinity" characterized by casual, even predatory sex and violence. Fatherless girls, like their brothers, tend to do less well in school and have greater difficulty in making the transition to adulthood. They are much more likely then girls who grew up with fathers to be young and unmarried when they first get pregnant.[27]

The predictable consequence is the inability of most disadvantaged children to develop the education and attitudes necessary for success at school and work. The concentration of these youngsters in central cities is partly responsible for the physical distancing from the mobile mainstream population escaping the seething social cauldron in the central cities.

The Moral Order and Suburban Secession

One outward manifestation of distancing from disaster area is the rise of the gated-community.

Secretary of Labor Robert Reich, suggests that the gated-community phenomenon is symptomatic of an isolated economic elite that is beginning to withdraw from society, in a new form of secession. Living in enclosed suburban communities protected by private security guards, shopping in secure suburban malls and working in suburbs, the upper income people are "resisting efforts to spread their taxes outside their own communities."[28]

Is this trend the result of a sudden access of selfishness? Reich points out that this elite is nervous, anxious about their jobs and fearful that their hard won security could vanish at the flash of a pink slip.[29]

In this state of mind, pleas to restructure the tax system or to engage in fair share programs for the sake of the inner city poor are not likely to resonate. We are the inheritors of generation of propaganda picturing government intervention efforts as ineffective, the sponsor of useless but costly social engineering and the poor as a feckless collection of losers. But perhaps the chief reason for secession is fear. Generations back when the gap between affluent and poor was at least as wide as it is in the 1990s, the poor were welcomed in middle and upper income households as day help or live-in servants and the high risk areas in cities were confined to the tenderloin districts.

Geographic secession reflects the fact that the affluent feel it prudent to move out of the line of fire. When the poor are perceived as menaces as well as losers and middle income people are insecure about the future, the abyss between the two nations grows wider and wider. In this broad context, telecommunications technology is a minor contributor to trends well under way before the computer moved into the marketplace.

Clockwork America?

It was the Koerner Commission that concluded, after its analysis of the courses of the race riots of the late 1960s that the U.S. was moving toward -- if it did not already have two societies, one mostly black and poor, the other mostly better off and white.[30]

Any realistic appraisal of the U.S. in the late-1990s would have to agree that the society is badly split but only partially along racial lines. A substantial portion of the black population, perhaps half, was working-class or middle-class with a few among the affluent. Meanwhile, underclass America includes many Latinos, native Americans, some Asians--and many rural and urban whites.

The class differences between the underclass and the mainstream seem to be widening. A gulf of behavior, attitudes, clothing, education, and speech patterns differentiates many of the poor from the rest of society. Indeed, the gap seems to be positively Dickensian, an American version of Anthony Burgess' *A Clockwork Orange*. Burgess' 1962 novel portrayed a society ravaged by predatory youngsters, living in degraded slums, violent, conscienceless, marked by a distinctive idiom and clothing, and traveling in dangerous gangs. He saw the police as powerless to stop gang warfare or halt incursions into middle-class households that are considered fair game by these urban wolf packs until there was a mainstream revenge: Brutal police, harsh incarceration, including coercive brainwashing. Burgess stops short of the "Yukon Solution" (distant relocation) proposed by an emigre Pole in the 1960s: An American gulag in underpopulated Alaska for social undesirables.

Are we well on our way toward this unhappy scenario? In the late-1990s there are heartfelt pleas to extend computers and computer training to low-income populations, suggestions which fail to take into account that this underclass lacks precomputer basics: literacy, numeracy and--too often overlooked-- standard speech skills. But the main thrust of public policy is a compound of exasperation and donor fatigue. Cutoffs of welfare benefits, tougher prisons, longer term incarceration, more gated-communities, a huge increase in private police and a proliferation of laws permitting ordinary citizens to carry concealed weapons reflect these attitudes. Moreover, the prospects are dim for any large scale "fair share" dispersion of the poor to the suburbs. At best, small numbers may be relocated on the nearly invisible Gautreaux Chicago model, but not enough to diminish the critical mass trapped in the seething hard-core slum.

With the fraying or outright rupture of personal contacts between mainstream and underclass--the end of compulsory military service contributed to this mutual insulation--the likelihood is for more demonization. Rap music, media depiction of sullen offenders apprehended for horrendous offenses and a string of personal anecdotes all feed the conviction that many of the very poor are dangerous barbarians who must be watched, kept away and if necessary, incarcerated to remove them from civilized society.

All in all, this is far from outright triage in which the mainstream cordons off the underclass and lets it molder in place. Indeed, based on past precedent, the future may hold more promise than now appears. Dicken's slum population survived, surmounting generations of terrible hardship. Similarly, based on past history, America's underclass will survive and eventually become transformed into a less destructive population. But in the foreseeable future, there is trouble ahead. Some problem-ridden areas will weather the storm, but in an age in which amenity wars will increasingly determine location, Uglyvilles will be losers, pulled down by crime, grime and misgovernment, by mediocre schools and above all, a perception of danger after crime rates most likely will recover from their temporary drop in the mid-1990s. The new technology is an added impetus to centrifugal demography. It gives more and more people the option of secession, relocating away from people and places they don't like.

The Moral Order: Rhetoric and Reality

Throughout most the 1960s to the mid-1990s, the moral values issue seemed to belong to conservatives, the upholders of heterosexual marriage, crusaders against gays, pornography and drug abuse (excepting nicotine and alcohol) vigilant against the menace of welfare cheats and defenders of the flag. Some of this fervor began to leach away in the 1990s, partly as a recognition that many rabid right-wing patriots were equally fervent in avoiding serious military exposure, prolonged monogamy and failed to resist financial temptation.

Basic contradictions face conservatives: the simultaneous demands for high quality public services, particularly public schools, criminal justice and defense and the conservatives' bedrock belief that reducing taxes for the well-off is the pathway to growth and happiness. Attempts to portray the rich as the 'investor class' desperately in need of capital gains tax relief and or a minimal flat tax to generate jobs for the poor seemed to have foundered on the fairness issue as well as the realization that most of the investment now comes from pension funds, insurance companies and other institutional sources rather from fat cats.

In addition to their other sins, the liberals were somehow held accountable for the loss of neighborhood values, for weakened private eleemosynary structures because the big government programs they mounted crowded out voluntary charities.

Fraying Ties

An integral element in the conservative tide sweeping through America since the late 1960s is the assumption that government intervention on behalf of the poor has not only failed but through welfare dependency and permissive go-easy-on-the-criminal attitudes has actually made things worse. This argument would be more persuasive were it not for the fact that the indices of decaying community standards began to show up by the late 1950s, before the Johnson Great Society legislation became a factor and that other nations with different cultures and legal systems are experiencing the same kind of problems. Racial tensions, unemployed youth, high rates of illegitimacy are a feature of much of Western Europe and are blossoming in the former Socialist nations. For example, it is difficult to identify a single strand of legal permissiveness as the cause of rising crime rates in the ethnic or religious conflicts that bedevil a number of European countries. Indeed, some liberal critics see capitalism as the villain.

> An alternative view lays the causes of social miseries and political floundering to the triumph of 'market worship': suggesting that the unfettered free market that has been most radically disruptive force in American life in the past generation, busting up neighborhoods and communities and eroding traditional standards of social life and personal conduct. It is the tyranny of the market that has destroyed the loyalty of corporations to their communities; customers to their neighborhood merchants; athletes to their local teams; teams to their cities.[31]

How much of the problem can be laid at the door of economics? No analysis can omit the decline of traditional organizations that once provided structure in a nation of joiners. Union membership is down; most fraternal organizations can't get enough young people to enroll; "bowling leagues have disappeared replaced by people who bowl alone."[32]

While slavish adherence to the bottom line is bad news for local corporate citizenship and for the "company men" who find their loyalty unrewarded and betrayed it does not account for the fact that the Masons can't attract young members or for declining enrollments in priestly vocations.

> The weakening of the social fabric is evident in the graying of volunteer associations. Older people dominate, few young people join. In 1935, in the midst of a depression, people who turned 18 were on the average members of 2.2 organizations such as the Rotary, Elks, Red Cross and PTA. This represented a rise from the Babbitt joiner level of the 1920's. In contrast, the level in the 1980s had dropped to just over 1.0.[33]

Why the steep decline? Robert Putnam lays the responsibility for the strange disappearance of civic America at the door of television. In 1950 less than one in ten U.S. households had TV sets but by 1960 nine in ten households were owners. Watching an average of even three hours of TV a day would absorb about 40 per cent of American's free time and TV viewing has increased over the years to an average of over six hours a day per household. No wonder Putnam feels that:

...each hour spent watching television is associated with less social trust and less group memberships. Television has 'privatized' leisure time because it comes at the expense of nearly every social activity outside the home, especially social gatherings and informal conversations.[34]

Is the situation as bad as Putnam suggests? Many organizations are flourishing despite the TV. Little Leagues, fundamentalist churches and private schools are only some of the associations that are growing in the midst of apathy and alienation. Certainly Putnam's depiction seems overdrawn.

For example, Michael Pollan, an exurban writer with an on-line office offers a more optimistic view:

> One of the great advantages of working at home is the added time with one's family it affords, once you realize how much day is left after you've lopped off those hours for commuting and water cooler schmoozing...
>
> The rap on bedroom communities has always been that a society of commuters with one foot in the city and the other in bed, had no time or energy left for local politics and community activities...I find that after a day spent in the solitude of my hut, the prospect of a PTA or zoning board meeting begins to look pretty exciting. A society of home-office workers could wind up actually revitalizing our communities. What might look like a formula for social atomization could prove precisely the opposite.[35]

Is there any compensation for the decline in active membership in many associations? We can speculate that the numbers of active members in extremist organizations like the Klan and the paranoid militias would be a lot higher were it not for the fact that sports and sitcoms provide their main evening entertainment for their potential membership rather than cross-burnings and other deviltry.

It's been suggested that the loss of traditional neighborliness based on physical proximity is partly compensated by the proliferation of virtual communities. In Howard Reingold's definition:

Virtual communities are social aggregations that emerge from the net when enough people carry on public discussions long enough with sufficient human feeling, to form webs of personal relationships in cyberspace.[36]

Reingold offers numerous anecdotes attesting to the close interpersonal connections that already exist and that are proliferating between people separated by great distances but tightly linked by the computer technology. These communities of common interest, friendships, clusters of interaction have little or nothing to do with traditional urbanism, i.e., geographically based communities. They are more akin to professional associations based on collegial ties than to the churches, PTO and the civic groups that provide the glue holding neighborhoods together.

One must agree with Graham and Marvin that virtual communities are not a full-blooded replica of the real thing.

...can telematics and virtual communities really substitute for the declining public spaces and sense of community in the real city? We have doubts about the extent to which convivial, face-to-face interaction and the public, democratic realm of urban places can be genuinely substituted by virtual communities. We are skeptical of the wilder claims for virtual communities. While it is easy to romanticise some long-lost 'public realm' or 'sense of community' in historic urban life, we argue that public interaction on streets and in public spaces offers much more than can ever be telemediated. Real face-to-face interaction, the chance encounter, the full exposure to the flux and clamor of urban life--in short, the richness of the human experience of place--will inevitably make a virtual community a very poor substitute.[37]

Our Town: But Not Our Plan

Concern over the weakening of linkages is much in evidence in the growing number of mergers, acquisitions, and business troubles that lead to plant closings. The age of home-grown manufacturing operations that form an integral element of the community seems to be ending. This offers one more reason for lamenting some of the changes that have taken place since the 1950s. Alan Ehrenhalt offers a good example:

> In 1895, David Lennox invented a new kind of steel furnace and set up a business in Marshalltown. Lennox Corporation became a solid source of respectable factory jobs that enabled generations of blue-collar families to enjoy the comforts of middle-class life. Its managers helped with countless local fairs, fund drives, and school buildings. Lennox probably could have improved its profit margins in the 1950s by moving to a place where labor was cheaper, but its loyalty was to Marshalltown...

> In 1993, Lennox...announced that it might have to close the Marshalltown plant altogether, not because the company was losing money or facing any sort of crisis, but just because the time had come to seek out the best opportunities. The fact that Marshalltown's very survival might depend on Lennox was of no consequence. "Strictly a business decision,"the company vice president said." [38]

What was totally unanticipated in the past generation was the hyperindividualism of the 1960s, the loss of family and community that makes the 1950s look awfully attractive in nostalgic retrospect, bathed in a golden glow of stability and social and rational expectations for a better future confirmed by recent experience.

To most observers, oratorical podium-thumping, fiery sermons, repressive legislation and outright mourning for past certainties seem to be little more than political froth. Squeezing the toothpaste back in the tube is not an exercise for realists. But does this mean that all we can look forward to is more of the same, only deeper stronger, and faster, more layoffs, more illegitimacy, more crime--and more impassioned speeches railing at the misbehavior of the lower classes?

This moral rebirth may be difficult to achieve in an era where personal moral probity is devalued as a prerequisite for political office or securing bank credit. Before Dun and Bradstreet, before the advent of tabloid free fire-zones on public personalities, character, or at least the illusion of rectitude, was a vital commodity. In the 1990s, personal failings seem to matter less and less when open homosexuals are reelected to office, adultery is ignored and bouts with the law, the bottle and the courts represent minor handicaps in business and politics. Michael Barone states, compared tothe 1940s and 1950s:

> We no longer have all these things that brought us together back then--community, public high schools, the draft, even the networks.

People go their own ways, watch their own cable channels, and live in completely different Americas.[39]

Funeral orations over the passing of happy times must be placed in historic perspective. Much of what passes for serious critiques of the decline in community, morality and civility in the 1990s in favor of fragmentation and polarization represents an exercise in false nostalgia, contrasting a mythical past against a troubled present. For three centuries America has been deeply riven by fissions between black and white, white and red, North and South and East and West. Sectional conflicts include the Tory-Rebel struggles in the Revolutionary War, and the Civil War followed by intermittent labor--management confrontations far worse than anything we see in the 1990s. Periodic racial outbreaks ranging from lynching to rioting, and the native vs immigrant battles go back into the nineteenth century when waves of immigrants from Ireland and southern and eastern Europe were the targets of nativist fury. And, it was not just the outright bigots who joined the opposition; labor unions fearful of low-wage immigrant competition and even liberals concerned about the decline of American society were prominent in the efforts to turn back the tide.

The issue of immigration has long split the liberal camp. Labor unions, for example, have consistently pressed for stricter immigration quotas as a means of protecting jobs from low-wage competitors. Richard Hofstader points out that Edward Ross, a progressive stalwart in the early part of the century, wrote a tract in 1914 that

> ...was unsparing with the currently most numerous immigrants from southern and eastern Europe. Immigration, he said, was good for the rich, the employing class, and a matter of indifference to the shortsighted professional classes with whom immigrants could not compete, but it was disastrous for native American workers. Immigrants were strike breakers and scabs, who lowered wage levels and reduced living standards toward their "pigsty mode of life," just as they brought social standards down to "their brawls and animal pleasures." They were unhygienic and alcoholic, they raised the rate of illiteracy and insanity; they lowered the tone of politics by introducing ethnic considerations. They threatened the position of women with their "coarse peasant philosophy of sex" and they debased the educational system with parochial schools and by selling their votes for protection and favors, increased the grip of the bosses

politics. They bred in such numbers that they threatened to "bastardize" American civilization.[40]

The Durable Moral Order

As a historic perspective to the allegations that traditional values are falling by the wayside, it is instructive to reexamine the litany of beliefs that Robert and Helen Lynd assembled in their chapter on "The Middletown Spirit." Written in the early 1930s, the book reflects an American ideology that the Lynds say remained basically unchanged from a decade earlier when the authors conducted their pioneering Middletown study. It seems clear that the basic credo had not been decisively altered by the awful experience of the Great Depression partly because most of the articulate business spokesmen had come through without substantial losses and partly because the victims and losers still remained believers. What are these adages and more specifically, what are the elements that continue to be relevant to America's secular moral order? Some of the cogent certainties include core beliefs:

- That change is slow, and abrupt changes or the speeding up of change in planning or revolutions is unnatural;
- that "in the end those who follow the middle course prove to be the wisest;
- that evils are inevitably present at many points but will largely cure themselves. "In the end all things will mend;"
- that within this process the individual must fend for himself and will in the long run get what he deserves, and therefore;
- that character, honesty, and ability will tell;
- that society should not coddle the man who does not work hard and save, for if a man does not "get on" it is his own fault;
- that Americans are the freest people in the world;
- that the American democratic form of government is the final and ideal form of government;
- that idleness and thriftlessness are only encouraged by taking charity;
- that business can run its own affairs best and the government should keep its hands off business;
- that competition is what makes progress and has made the United States great,
- that a chance to grow rich is necessary to keep initiative alive;
- that ordinarily any man willing to work can get a job;
- that "any man who is willing to work hard and to be thrifty and improve in his spare time can get to the top;"

- that the rich are, by and large, more intelligent and industrious than the poor;
- that taxes should be kept down;
- Middletown is against the reverse of the things it is for.

These may be summarized by saying that Middletown is against:
- centralized government, bureaucracy, large-scale planning by government,
- anything that curtails money making,
- anybody who criticizes any fundamental institution.[41]

I have eliminated from this credo notions that have fundamentally changed in sixty years, particularly the casual religious and ethnic denigration that remained a fundamental part of American life until Hitler gave bigotry a bad name (that is, until after World War II). Most of these bedrock beliefs were as prevalent in the nation as in Middletown and for the most part, are still very much alive in the 1990s and are likely to remain so into the foreseeable future. Certainly appeals to this fundamental belief system have been a basic element in American political life including the 1996 presidential election with its attacks on "extremists" appeals to the "vital center", promises to revamp welfare dependency to reward competitive free enterprise, to reduce tax burdens and pledges to downsize big government. Compared to the traumas this credo has weathered--wars, popular and otherwise, depressions, riots, rural depopulation, racial turmoil, waves of crime and drug usage--the shocks inherent in the diffusion of telecommunication technology are mere flea bites to be taken in stride like predecessor inventions in communications technology, the telephone, movies and the automobile.

Ehrenhalt offers, by way of caution and consolation two historic examples that prove that an era of cynicism and moral decline can be superseded by structure and order. In the U.S. he offers the example of the Jazz Era, the Roaring '20s when skirts went up, moral standards went down, the KKK was resurgent and capitalism rapacious and unchecked. The prevailing nostalgia for social conservatives in the 1920s was for the happy pre-World War I era, when small towns provided a highly structured living environment ruled by strict moral codes and unquestioned authority. Disillusionment with politics and rampant Babbitry seemed fixed and irreversible in the 1920s but it was not to be.

As Ehrenhalt demonstrates, the Great Depression and the Second World War, back-to-back restored a sense of common purpose and social cohesion, of faith in civil and military authority that had seemed lost forever in the 1920s Lost Generation. To suggestions that similar life-threatening crises are not on hand to bring an atomized nation together in the early part of the next century, Ehrenhalt offers an historic example when morality was restored without national trauma.

England in the 1820s was ruled by a profligate king and quarreling queen, beset by political corruption, loss of faith in the established church, an addiction to alcohol and in some quarters, to opium. There were familiar laments for the lost stability of the good old days before the French Revolution, fear that class harmony was gone forever. The ruling classes not only were setting a bad example with their heavy, gambling, drinking and sexual misbehavior but fearing revolution and discord, ruthlessly crushed every manifestation of trade unionism and hanged or transported even the pettiest of criminals.

But, a generation later all was transformed. The artistic but wicked Regency gave way to stodgy Victoria and Albert, Victorianism replaced profligacy--a new era marked by Christmas trees, family values and a reformed religious establishment. All this without a big war or major depression.

Can the historical pendulum swing back in the U.S. in the absence of a major, unifying crisis? Ehrenhalt raises the prospect that the generation of the 1960s might be tempted to move in the direction of order and stability as they revolt in horror and disgust against family and community breakdown.

Trying to Revive The Sense of Community

At the community level, calls for "tough love" and "end-welfare-as-we-know it" are often accompanied by assertions that the removal of governmental safety nets will bring back a return to pre-New Deal community-based, neighborly charity: help for the deserving poor and a resuscitation of shame-for-the-shiftless, the feckless, undeserving parasites who teach, by example, their offspring to mooch on the working population.

Some of the calls for civic action came from the grassroots, from neighborhoods threatened by crime or by job losses. There are dozens of inspirational stories that offer evidence that local action can made a difference.[42]

While the reasons for the worry are clear, the causes are variously ascribed to:
- The impact of television in creating a passive, cynical public and cutting into evening and weekend volunteering time. (Occasionally air conditioning which keeps people indoors is also named as a culprit);
- the generational revolt of the 1960s which challenged authority and conventional morality and introduced a bewildering variety of life choices; among them is the proliferation of full-time working mothers with little time to devote to her children;
- the impact of materialistic capitalism that has made two income families a necessity and has weakened or dissolved the ties between corporation and employees, corporation and community and labor union solidarity in an everyone-for-himself workplace;
- the attack on government and politicians from left and right over the past six decades.

The 1960s and 1970s were a watershed with the corrosive loss of faith in the architects of an increasingly unpopular war in Vietnam and the simultaneous Civil Rights turmoil. But there is in the background the six decades of unremitting attacks on government as costly bumbler and vicious tyrant (e.g., the IRS) found in many issues of *Readers Digest* and echoed in right-of-center publications. However, disenchantment with the private sector surfaced in the 1990s with the spectacle of richly rewarded management laying off hundreds of thousands of employees in reengineering and downsizing sweeps, rationalized as for necessary for corporate survival in a globally competitive world.

Alan Ehrenhalt in *The Lost City* contrasts three Chicago neighborhoods of the 1950s; one working-class white, one black, and one white suburban with the turbulent scene of the mid-1990s. As he points out, crime rates were much, much lower, most people abided by the rules and respected authority while coexisting with crooked municipal government and fraudulent unions. Many were active members of churches and fraternal associations.[43]

In the 1990s suburban residence is generally associated with moderate amounts of neighboring--occasional backyard barbecues, interfamily visits to swimming pools and skating rinks, birthday parties and Fourth of July outings. By contrast, in the 1950s there was, as Ehrenhalt points out, an intense pressure to join and participate, to conform, to pry and to entertain one's friends, relations and neighbors.

People tended to stay married in the 1950s to their spouses, to their political machines, to their baseball teams. Corporations also stayed married to the communities in which they originated.

In contrast to past decades, Eric Uslaner sees Americans becoming a nation of Ebenezer Scrooges, "dour, distrustful, sour and disconnected." In his view, Americans vote less often, join fewer political and social groups, volunteer less, give a smaller share of our gross national income to charity, and don't socialize as much with our neighbors as we did. What began as a sizable cynical minority in the 1960s (42 percent) became the majority opinion in the mid-1990s when full two-thirds of Americans fall in the "distrustful" category.

Uslaner warns of the unique element in the emerging generation of mistrusters. In earlier times the disaffected joined together to change the system. Today they are pessimists who don't get involved. They "just stay home." [44]

Reality and Created Reality

It has become something a commonplace to hear diatribes against the communities and shopping malls, the restaurants and civic centers that have deliberately and carefully planned environments that stimulate, sterilize and romanticize untidy reality. The trend is most pronounced in theme parks but is echoed in the creation of streams and forests within malls and hotel atriums and to replications of a glorified Old West, Old Mexico and their half-imaginary stage sets in downtown civic centers, in shopping malls, and often in airports.[45]

There is no secret about the reason for the creation of "idealized reality," the colorful description of MCA developments sterilized recreation of Hollywood Boulevard and Venice Beach on the grounds of Universal Studios theme park in the San Fernando Valley. Critics may denigrate it as an attempt to "censor the realities of urban life" but as long as the public pays there will be more of the same.

Orlando, Florida, eastern home of Disney includes a dazzling variety of tamed exotica- medieval dinner theaters, a rain forest cafe, African safaris, a China theme park including a miniature Great Wall, Mayan Temples, Egyptian pyramids, simulations of Paris and Rome. It seems clear that MCA and Disney simply represent the latest manifestation of an old trend. In the early industrial age in the nineteenth century, wealthy Englishmen created artificial landscapes, and picturesque ruins on their properties in contrast to the muck and grime of the coal mines and industrial cities that made them rich.

Theme parks are only one part of the reversion to an idealized past. The neo-traditional town planning that seemed to grow in popularity through the 1980s and 1990s is a prime example--and for excellent reason. As one analyst sees it, while detractors say they are artificial and over controlled stage sets, a neo-traditional community offers tangible attractions, compared to existing market alternatives. In the case of the Disney company's Celebration where housing sells about $100 a square foot versus nearby Hunter's Creek where housing runs about $75 a square foot:

> In Hunter's Creek, a community of more than 10,000 people, homes are clustered into several gated-neighborhoods. The homes within each neighborhood look vaguely alike, but home buyers of different incomes are segregated. One neighborhood has houses for $110,000 to $150,000; another has houses from $200,000 to $220,000; another has homes from $230,000 and up.
>
> From any of these neighborhoods it's a car ride to the "shopping village," basically a big strip mall with a Publix, a Kmart, a Subway, a Radio Shack, a Little Caesar's and several other stores.
>
> The geographical heart of Hunter's Creek is the busy intersection where two four-lane roadways meet at the "shopping village." Cars whiz by. Crossing the street from one sidewalk to another means traversing more than 150 feet of grassy berms, a traffic island and several lanes of traffic. There's no shade. No one walked in the "town center."
>
> Now visit Celebration. The geographical heart of the community is the quaint town center, which as it happens, draws on more than mere nostalgia.
>
> The town center's focus is a promenade that winds around the pond. There are ample benches at waterside, some shaded by graceful wooden structures, and they transform the waterfront into a gracious public living room.

From there, walk up Market Street, the downtown spine. It's lined with shops. Above shops there are apartments.[46]

In short, in contrast to social distancing at least some of the physical fragmentation associated with postwar land use planning is reversible. There is a sizable market for many people willing to pay for a community-oriented alternative.

The High Road to Psychological Breakdowns?

Virtual reality, like all forms of communication, can be used as a powerful teaching tool. For example, it could be a prime painless, method of imparting living history, strolling the streets of ancient Rome or Athens, witnessing the reality--eventually including the smells--of Shakespeare's London and the antebellum south, accompanying, briefly the turn-of-the century immigrants and recreating realistic snippets of experience in the Great Depression and the Great Wars.

From a planning standpoint, the perpetual adult education seminar that constitutes much of a planner's task could be enlivened by lessons-by-analogy, having one's audience take a guided tour to communities that have done a good job of coping with their problems and those that have failed. In effect, this is PBS and network public service multiplied by a factor of one hundred.

The possibilities for diffusion of learning are staggering. We can have instant, total immersion to absorb languages, to train in a wide range of vocations, to grasp difficult scientific, biological and medical concepts and techniques. We will be able to zero-in as close observers on the scene at historic dramas, revolutions, scientific discovery, explorations. From a pragmatic standpoint, planners will be able to teach by analogy, conducting standpoint, planners will be able to teach by analogy, conducting commissioners and legislators to communities where one or another alternatives can be viewed in action. Intense religious experiences can be created on demand, along with highly persuasive political and commercial advertising. but there are darker possibilities. As Clifford Stoll sees it:

> The computer is a remarkably different tool--one which can turn kids into reactive zombies, adults into frustrated bumblers...The network

presents an unreal world where you can appear to be anyone you wish. Adopt a friendly persona or that of a grinch. But you aren't who you pretend to be. Inside, you're still you... You have nothing physical invested. No matter where you appear to be you're always in the same place. No matter how dangerous the situation seems, you're always safer. There's no need to tolerate the imperfections of real people.[47]

Is there any way of deflecting, averting or stopping in its tracks a new technology that may offer a short cut to dementia? Can we control a tempting pathway for stunting social growth and a danger to the nation's mental stability. The short answer is no. After all, we're adventurous Americans. Suppose three generations back we were offered a devil's bargain. Freedom to work in more places, freedom to travel and shop, freedom for romance, free from prying eyes. The cost: a million dead in highway accidents, millions more injured. Would we have said "no thanks, we'll stick to the streetcar;" confining, limited but very safe. Certainly not. And so for better or worse we will risk new nightmares in a new century. What will the new technology offer?

At a mundane level, operators can rewrite history, serve heroically, and painlessly with Roman Legions, Napoleon's Grand Army, Nelson's fleet or Stonewall Jackson's corps. One can be a virtual business tycoon, inventor, actor, or sage. And with the help of animation and squadrons of writer/compilers we can chat with Jesus, Moses, Confucius, Thomas Aquinas, Montaigne and Erasmus--all of them, fluent in contemporary American English.

We already have a generation of a social computer nerds as lacking in social graces as their predecessor "bookworms," of earlier generations. What virtual reality--the next step up in computer technology--offers is more enticing. Assuming that introversion-extroversion can be measured on a scale of 1-10, we are likely to widen. the spectrum of addicts from perhaps the ones and twos to perhaps the threes and fours. Given a spell of hard luck or bad experience in the real world, the temptation will be to retreat to a safe universe where one will never, never, be a loser. Even more disquieting is the capture of teenagers who rarely leave their telecommunications haven and are thereby unfit for a real world of successes and failures, short of ability to cope with bad times, rejections, and disappointments in work and personal relationships.

In previous generations, frustration at home led to vows of lighting out for the territory in Mark Twain's era, joining a convent, a cult, the military, taking to the bottle or drugs, or diving into an alluring hobby. Virtual reality makes retreat easy and safe, almost effortless. It takes the fear out of withdrawal from reality.

We can foresee some time early in the coming century, a growing, hidden population of reclusive attic people. It has been a tradition among affluent WASPs, particularly in the South, to tuck daft Uncle Wilbur and fey Aunt May in a back bedroom, she to pursue genealogy and memories, he to write his immense, disjointed novel. If, the alternative world of vicarious reality nips socializing early in life or even at mid-career we can proliferate Uncle Wilburs and Aunt Mays by the tens of thousands. Harmless, and useless, cut off real life they will be lost to society as surely as if they were heavily sedated, human vegetables in a mental institution.

Indeed, Howard Reingold, sees the possibility that many people may largely abandon the "real world" in the future, preferring to live in the fictitious worlds created by the entertainment industry. If this is the shape of the future we can expect a boom in the mental health industry which will try to wean Cyber junkies way from nirvana to a real world that has its downs as well as its ups. As I suggest, troubled teenagers may be particularly prone to a retreat from a real world of possible snubs and insecurities, of social awkwardness, shyness and self doubts.

> In the years ahead we will live increasingly in fictions: we will turn on our virtual-reality systems and lie back, experiencing heavenly pleasures of sight and sound in a snug electronic nest. The real, world will be almost totally blotted out from our experience." One participant in the present world of computer interaction offers a glimpse into the future.

> The latest computer communications media seem to dissolve boundaries of identity... I have three or four personae myself, in different virtual communities around. ... new identities, false identities, multiple identities, exploratory identities, are available in different manifestations of the medium.[48]

The possibility of psychological problems when virtual reality is in everyone's home waiting in the wings is frightening. We can foresee addicts who create false identities much more fascinating than their real selves. What happens when one or more of the personae may be too tempting to let go? And what happens when the keyboard is supplemented or supplanted by visuals? Will the players adopt masks, costumes or create visual imagery in place of the real self?

The second escape, retreat into cyberspace may also have its hazards, (aside from the dangers to mental health.) Sterling underscores the seedy side of the Internet, leaving aside the menace of privacy invaders, eavesdroppers and virus-mongers. We can add dangerous cults, sexual predators and con artists to this list.

Will virtual reality shield its practitioners (or more precisely its addicts) from the messy real world of personal relationships that may entail risks? As a Cyber columnist puts it:

> "It's easier to be erudite and witty at the great cyber cocktail party where you can edit your comments before hitting the send button than it is fumbling for words at the real thing while trying to balance your white wine and cheese ball plate in one hand as you push our glasses up with your finger. On the Internet no one can tell you have a piece of spinach stuck on your front tooth."[49] N.B.(This is likely to change when two-way video becomes a reality. Careful grooming will be de rigueur.)

Examples of computer-addiction began to surface in the 1980s; by the mid-1990s there were support groups for people out of control. "Caught in the Web" is a group formed at the University of Maryland to counsel students spending too much time on computers. At the Massachusetts Institute of Technology addicts unable to break the habit on their own--playing computer games--can request the University to "lock the refrigerator"; i.e.,deny access to the campus terminals when the urge to sign-on mounts. Half the freshman dropouts at Alfred University in New York had been logging marathon, late-night Internet time.

A Columbia University official: "We're seeing them (i.e., students) really drift off into this world at the expense of practically everything else." It appears that for students having trouble establishing ties at a large university its becoming a tremendous escape from the pressures of college life.[50]

When you are putting in seventy or eighty hours a week on your fantasy character, you don't have much time left for a healthy social life.

One key to technologic impact is diffusion. The halfway point on the ownership of a home telephone was not reached until the mid-1940s, 70 years after the device was invented. It my be remembered that the *Literary Digest* poll which predicted a Landon victory over Roosevelt in 1936 was an utter failure because it relied on a telephone survey at a time when many poor voters were not subscribers. A similar poll in the early 1990s based on personal computer ownership might be off-the-mark by a similar margin: The ownership level was just under a third in 1994.[51]

As was the case with television, the diffusion of computers is greatly stimulated by the fact that the device is an entertainment as well as an educational and business tool. In the late 1990s 83 per cent of adult computer owners reported using their home computers for personal use compared to 67 per cent for job use.[52]

Given the popularity of fantasy including games, pornography, and chat rooms, we can foresee a huge increase in the numbers of people who have the means and inclination to join the action. One might add, on a personal note, an enormous market for parents and grandparents who wish to stay in visual contact with offspring as soon as two-way camera technology becomes simple and affordable.

A special boost to usage will be low-cost virtual reality when this technology becomes widely owned in coming decades.

Some analysts forecast that the virtual reality market, only $90 million in sales in 1995, will reach $6 billion by the year 2000. And why not? Games are leading the way. Armed with headsets and gloves, fun seekers can play intricate video games. But the potential has great appeal to those of a more spiritual bent. Virtual reality can be a marketing tool for missionaries to developing countries. How? Programs maybe used to recreate bible stories as a powerful tool for preaching to nonreaders. "You could let people interact with a virtual Jesus."[53]

The confirmed couch potato, driven by obsessive behavior patterns, disheartened by forays into reality, will exhibit an outright addiction to a protected universe. Threatening interaction can be erased at the flick of a finger. True, the healthy personality can flourish with a variety of new friends and like the epistolary comrades of old, people can grow close over distances and over the years. The danger is the seduction of the stunted, friendless adolescent, the rejected adult, the lonely, the unhappy who find screen life infinitely cozier than a real world of occasional hard knocks, periodic failure, stretches of boredom and indifferent and occasionally unsettling interpersonal contacts.

We have another extrapolation of the technological future:

> People will no longer visit foreign places; foreign places will visit people, in the form of Disney's Multisensory Package Tours. These tours will enable you to visit not merely other places but also other times. Disney's most popular product will enable the consumer to star in his own fictions--movies, music videos, novels, etc. All the most popular fictions will be set in the distant past. The future will have lost all its glamour, as it will be impossible to imagine it as anything but more of the present. *The Kugelmass Episode*--Woody Allen's short story about a Brooklyn man who, seeking to slake his midlife crisis, falls in love with Emma Bovary and is thrust into the text of Flaubet's novel,will have become reality.[54]

In assessing the future of virtual reality, it is useful to consider some of the foreseeable limitations. Virtual it may be, but taste and smell are missing and sight is not enough like the real thing. Over and above the dangers of theft of credit card numbers, online shopping where net users can browse through catalogues and stroll down aisles to view the goods has not taken off into the wild blue yonder.[55]

The arrival of the body suit will bring reality closer, (i.e., touch) but other senses will remain deprived until a generation or two in the future when the telecommunications can deliver user-friendly taste and fragrance.

By the early part of the next century we are due for virtual reality, not only the helmet and gloves now on the market but full body suits. This opens up starting possibilities, a universe of virtual reality personas, handsomer and healthier than the couch potato operators, a world of action-without-rejection, unfettered by age, illness, acne or utter social ineptitude. Even without virtual reality we have back page advertisements in computer magazines offering erotic web sites--"The Web's only on-line brothel," "Beautiful girls at your command," "One-on-one live encounters." Some analysts speculate that porn pioneers will led the way: the sexually frustrated will willingly pay high prices for high end, costly technology. The primary market for virtual porn is heterosexual men, in their twenties, thirties and forties, but there is an emerging market for women and homosexuals.[56]

The Machine Stops?

From time to time in the past two decades major and minor power outages have served as reminders that an advanced economy rests on fragile foundations. We have also experienced terrorist bombings in Oklahoma City and New York City that prove that large populations can be paralyzed by a few ideological loonies possessed of modest technical skills.

At the turn of the century E.M. Foster wrote a short story entitled *The Machine Stops*.[57] He pictured a world in which people lived as virtual recluses, fed, entertained and given clean air by the Machine, a world of total dependency on technology. And when one day, the machine stops there is total panic, total helplessness.

Are we due for a similar nasty encounter with reality? Will Foster prove prophetic and one day a nation of flabby technology-ridden slugs will find that years as a couch potato is poor preparation for the day when the computer is down and the back-up systems can't cope?

Probably not that far, not that bad--perhaps five to ten per cent of the population a generation or so in the future will be paralyzed when they lose their fix, incapable of coping when their artificial lifeline is cut: Not a happy prospect but a realistic one.

Other prominent authors have also ventured into futurology. George Orwell's *1984* depicted a dystopia in which telecommunications technology (i.e., visual surveillance screens) offers intrusive agencies the ability to exercise a suffocating degree of control over subject populations.[58] The kind of surveillance now found only in maximum security prisons is allied to advanced propaganda techniques to enable totalitarian Big Brother to manipulate, scrutinize and oppress the citizenry stamping out every overt sign of rebellion.

To summarize: We are less than a decade away from a major advance in communications technology that holds enormous promise for education--and serious risks for a substantial part of the population. We can foresee a radiant future for learning along with a golden age for mental health therapists.

1. Patricia L. Mokhtarian, "A Synthetic Approach to Estimating the Impacts of Telecommuting on Travel," paper prepared for the TIMP conference, Williamsburg, VA, October 27-30, 1996, 8.
2. Ibid., 15, 16.
3. Ibid., 4
4. "The New Workplace", *Business Week,* April 29, 1996, 109.
5. Ibid.
6. Cornish, op.cit., 6.
7. Joseph F. Coates, "The Highly Probable Future: 83 Assumptions about the Year 2025," *World Future Society,* (Bethesda, MD:1994).
8. Robert Lynd, *Middletown in Transition,* (London: Constable, 1937).
9. HBO's, "Cosmic Slop," an original 1996 drama,had the final solution to America's race problem accomplished at the instigation of an alien culture arriving in an omnipotent spaceship. The reaction of the white population ranged from indignation and passivity to relief and collaboration.
10. Sidney Wilhelm, *Who Needs The Negro,* (Cambridge: Schenkman, 1970)
11. Bruce Sterling, "The Virtual City," Rice Design Alliance Conference, Houston Texas, March 2, 1994
12. Lester Thurow, "Their World Might Crumble," *The New York Times Magazine,* November 19, 1995, 78-79.
13. Jeremy Rifkin, "Civil Society in the Information Age," *The Nation,* February 26, 1996, 11.
15. Wilt Ferry, Center for Democratic Studies quoted in Edward D. Kalacheck, "Automation and Full Employment," *Trans-Action,* March 1967, 24.
15. John T. Dunlop, ed., *Automation and Technological Change, The American Assembly,* (New York: Spectrum, 1962), 1.
16. Ibid., 110
17. Ibid., 29.
18. Gerald K. O'Neill, 2081: *A Hopeful View of the Human Future* (New York: Simon and Schuster, 1981), 52.
19. Robert O. Metzger and Mary Ann Von Glinow, "Off-Site Workers: At Home and Abroad," *California Management Review,* Spring 1988,110.
20. James K. Glassman, Jobs: The (Woe Is) Me Generation," *The Washington Post,* March 19, 1996, A 17.
21. For sheer gloom, see Kirkpatrick Sale, *Rebels Against the Future: The Luddites and their War on the Industrial Revolution-- Lessons for the Commuter Age,* (New York: Addison Wesley), 1995.
22. Joseph Nocera, "Living with Layoffs," *Fortune,* April 1, 1996, 71.
23. Graham and Marvin, op.cit., 230-231.
24. Nicholas Eberstadt, *Society,* January/February 1996.

2 5. Cited in William Raspberry, "Out of Wedlock, Out of Luck," *The Washington Post*, February 25, 1994, A 21.
2 6. "The Family," *The Economist*, September 9th-15th, 1995, 26
2 7. Ibid., 29.
2 8. Frank Swoboda,"Reich Voices Concern over Growing Economic Elitism," *The Washington Post*, December 26, 1995, E1,E2.
2 9. Ibid.
3 0. National Commission on Civil Disorders (New York:Bantam,1968) Koerner was the Chairman of the Commission.
3 1. Alan Ehrenhalt, " No Conservatives Need Apply", *The New York Times*, November 19, 1995, 5.
3 3. Ibid.
3 3. Robert Putnam, "Bowling Alone is Back," *The Washington Post*, April 10. 1996, A-19.
3 4. Ibid.
3 5. Michael Pollan, "Living at the Office," *The New York Times*, March 14, 1997, A33.
3 6. Howard Reingold, *The Virtual Community: Homesteading on the Electronic Frontier*, (Addison-Wesley Readings:1993), 5
3 7. Graham and Marvin, op.cit., 231
3 8. Alan Ehrenhalt, *The Lost City*, (New York: Basic Books, 1995), 11-12.
3 9. Michael Barone, cited in Diana McLellan, "Mr. America," *Washingtonian*, September 1996, 43.
4 0. Richard Hofstader, *The Age of Reform* , (New York:Vintage, 1955), 179-180.
4 1. Robert S. Lynd and Helen Merrell Lynd, *Middletown in Transition: a Study in Cultural Conflicts*, (London: Constable, 1937), 406-418.
4 2. Herbert Wray, "The Revival of Civic Life," *U.S. News and World Report*, January 29, 1996, 63-67.
4 3. Ehrenhalt, op.cit.
4 4. Eric Uslaner, "Dire Conseuences When Society's Defectors Become a Majority, *Outlook*, The University of Maryland, Vol.10, No.24, April 4, 1996, 1,5.
4 5. Ada Louise Huxtable, "Living with the Fake and Liking It,*The New York Time*, Section 2, 1, 40.
4 6. Robert Whoriskey, "Dealing on Nostalgia," *San Jose Mercury News*, January 4, 1997, 37.
4 7. Clifford Stoll, *Silicon Snake Oil* (New York: Doubleday,1995), 45, 57-58.
4 8. Reingold, op.cit. , 147-148.
4 9. Larry R. Moffitt, "The On-Ramp," *The Washington Times*, September 4, 1995, Section C16.

5 0. Rene Sanchez, "Hooked On-Line and Sinking," *The Washington Post*, May 22, 1996, 1, 17

5 1. "Science and Technology," in *The Macmillan Visual Almanac*, (New York: Blackbirch Press, 1996), 521,522,526.

5 2. Ibid., 527.

5 3. Barnaby J. Feder, "Selling Virtual Reality, in Indiana," *The New York Times*, August 7, 1995, 32.

5 4. Michael Lewis, "Future Stock", *The New York Times Magazine*, September 25, 1996, 85

5 5. Winston Fletcher, "Home Shopping Take a Tumble from its Trolley," *Financial Times*, January 2, 1996.

5 6. Michael Marriott, "Virtual Porn: Ultimate Tease, *The New York Times*, October 4, 1995, C4.

5 7. E.M. Forster, "The Machine Stops," *The Eternal Moment and Other Stories*, (Oxford: 1904).

5 8. George Orwell, *1984*, (New York: Harcourt Brace,1983).

In Conclusion

It seems clear that in the short to medium term--say ten to twenty years--the principal impact of the emerging telecommunications technology on urban demography will be more of the same, only faster and deeper: more growth in recreation areas and selected college towns, accelerated expansion in upscale suburbs and attractive outlying communities.

On the other side of the balance sheet, we can expect more trouble in many central cities and most poor inner suburbs to go downhill even faster. Some will weather the storm but in an age in which amenity wars will increasingly determine location, Uglyvilles will be losers, pulled down by crime and misgovernment, mediocre schools and above all, a perception of danger when crime rates will likely recover from their drop in the mid-1990s.

In my opinion, telecommunications technology will enable middle and upper income people to complete the process of secession from problem places and problem people. This trend has been in full bloom for at least half a century. As Michael Conzen reminds us, as early as the 1970s in a well-to-do Philadelphia suburb, only one quarter of the residents visited downtown more than once a month and nearly one-fifth traveled there less than twice a year.... In general, many suburban rings are approaching a kind of institutional completeness that spells autonomy for day-to-day living. [1]

What was evident in the early 1970s is even more pronounced in the late 1990s and in another quarter of a century is likely to result in a social chasm between the well-off and those unable to escape the grime, crime and other stigmata associated with central city and inner suburban places and people.

If the general directions are visible enough in terms of emerging urban form, they are clouded when it comes to the content of urban life. On the one hand there is the prospect of dazzling new dimensions in education, in travel, a world of possibilities as we move into virtual reality in the early years of the next century. But there is a downside. In general, the prospect seems to lie in the direction of social fragmentation perhaps to the point of atomization with poorly understood virtual communities replace the civic associations and proximity neighborhoods that enrich community life. And there are darker possibilities: more illness, mental and physical, a widening gulf between knowledge-rich and badly educated poor, and the potential for large scale turbulence in the labor market with the threat of more long term unemployment for the ill-prepared. There is a distinct possibility of unnerving societal stress in a generation of rapid and accelerating technological change, a turbulence that leaves politics and government limping far behind. We can anticipate more open meritocracy with a weakening of racial, religious and class barriers in favor of free movement up the ladder--with the ever present risk of a precipitous fall. Security may be hard to find and harder to keep. As the ancient Chinese curse would have it, we are condemned to live interesting times. Bon voyage to us all.

None of this suggests that doom--or great expectations are a foregone conclusion. for the central cities there is hope in the strengthening of niche markets and in the prospect and reality of partial population replacement leading to a smaller proportion of dysfunctional native-born, and an influx of industrious, frugal immigrants. The time of trial of gateway cities may be a generation postponed: will immigrants remain in reborn cities or will their loser offspring take to crime and remain and their mainstream progeny depart for the suburbs?

At a regional level, we can follow the planner's consensus and emulate Portland or follow the path of inertia to waste, endless urban accretion without centers.

Most Easterners and Midwesterners are confronted with the intractabilities and blighted hopes of major and minor cities, Portland, Oregon seems an anomaly: attractive and safe, growing, thriving and compact, a trip down memory lane to a distant half century past when eastern cities were better, more neighborly places. As James Howard

Kunstler asks:

> Could this be America? A vibrant downtown, the sidewalks full of purposeful-looking citizens, clean, well-care-for buildings, electric trolleys, shop fronts with nice things on display, water fountains that work, cops on bikes, greenery everywhere?
> Portland, a city of 429,000, a little bigger than Pittsburgh and a bit smaller than Atlanta, seems to defy the forces that elsewhere drag American urban life into squalor and chaos.[2]

What is responsible for this happy exception to the rule. (There are others, perhaps not as outstandingly successful.) Kunstler sees the causation as a combination of climate, a geographic setting mandating compactness. And there is a vigorous planning tradition supported by an ecologically oriented, sophisticated public. Zoning, and height restrictions, incentives for downtown housing, a light rail (i.e., streetcars) serving city-to-suburb travelers, a "parking lid" on downtown parking spaces.

Is Portland doable in the east? Can it be cloned, modeled, replicated or at least approximated? At first glance -- no. Portland's population is unusual, better educated, "greener," less racially conflicted than the population components of most other cities. And there is the spilt milk factor. Once dispersion, massive rot, misgovernment and mistrust take root it is difficult to effect a turnaround. But difficult is not impossible. We can at least hope that ten or twenty years hence harsh experience, higher educational standards and new people (who will be partly responsible for turning over a new leaf) will turn cities around. Expectation would be too strong a word. Possibility is more accurate.

In an age of litigation this entails a broad political and cultural consensus lasting a generation. The top down elitist planning so effective in earlier eras such as Haussman's transformation of Paris in the Second Empire, Burnham's major City Beautiful impact on Chicago's waterfront at the turn of the century, huge constructions in New York--are no longer achievable. There has to be broad agreement and patient, consistent support, year after year. Portland shows that it can be done.

Equally significant are the neo-traditional developments (include the Disney Celebration among their number) that show a sizable fraction of the public is willing to pay premium prices for new housing in communities that look and feel towns rather than subdivisions. This I feel will be the wave of the future, particularly if Celebration fulfills its promise.

We can anticipate residual differences between liberals and conservatives in their response to lingering problems of the losers, the people that remain in the Uglyvilles. Liberals will deplore, from a distance, from homes in upscale suburbs, exurbs, and college towns, the miseries of those left behind and will be willing to vote for remedial government programs. Conservatives will express loathing, contempt and pity, from a distance (more of them in outer suburbs and recreation areas) and will call on the poor to reform and leave poverty behind them by virtue of their own efforts. In either case, physical interaction will be minimal.

Social Consequences: Polarization/Fragmentation

It is becoming possible to pursue a career in business, professional or technical occupations and live one's life without having any but the most fleeting contact with poor people. And the poorest may stay poor longer. An underclass lacking literacy and numeracy may be counted out of the mainstream when the entrance fee is raised to include teleworking skills.

Finally, there is on the near horizon the prospect of two technological advances. The first is universal two-way video communication. This is likely to reduce further the need for personal, face-to-face contact. But the giant leap toward depersonalization may come with virtual reality: widespread use of headsets and gloves soon, body suits down the road. We can foresee a huge increase in couch potato dropouts, heroes in private universes where rejection never happens and victory in love and war is a certainty. Experience with Dungeons and Dragons, that frightening teenage addiction, is only a foretaste of things to come. One need hardly say that therapy counseling will be one of the growth industries of the next century.

These grim possibilities must be viewed in context. I have underscored the negatives. The new technology also opens up dazzling possibilities for education and training, for travel, for new experiences and enhancement of culture and learning. In short, on balance, the consequences are likely to be mixed as is the case with radio, television and movies.

In the context of fraying social ties across the economic spectrum, it is useful to restate the case for the institution of compulsory national service, the benefits: personal experience at an impressionable age--the late teens--with

- arbitrary authority,
- exposure to persons of vastly different backgrounds,
- performance of useful work, and
- insistence on a demanding physical regimen: exercise, diet and medical care.

Most important perhaps is less demonization of stigmatized populations in an age when personal contacts are being replaced by media shorthand, by caricatures of reality and conspiracy theories. America will never be a classless society but it can move in the direction of easing the corrosive suspicions and disgust that left unchecked, will add a dose of poison to political and social relations.

Along the way, we can anticipate belated epiphanies on the part of political and business leaders hitherto blind or indifferent to esthetic considerations. Like the converts to historic preservation who discovered its charms when old buildings began to be priced like irreplaceable collectibles, civic leaders will emerge as tub thumpers for policies they once derided as prissy, antibusiness or irrelevant: programs such as urban design, billboard control, tough environmental regulation. And like many who attempted to replicate overnight North Carolina's Research Triangle in the '80s and '90s when their local industries, dehydrated or collapsed, these belated recruits to the amenity wars will discover that areas that look bad got that way over time and that repairs need money, patience and sustained effort.

If we are looking for persuasive prophecies of things to come we can turn to Bill Gates of Microsoft.

> ...Gates comes up with a long list of ways in which a future broadband information network will provide, 'The opportunity for people to disperse more than they can now.' There will be sophisticated video conferences, remote collaboration and a definite rise in in telecommuting. He also thinks companies will keep a smaller workforce to concentrate on their core competencies, and farm out

less essential work to consultants, who will be able to live pretty much where they want.[3]

Before succumbing to despondency in the face of the enormous challenges confronting us in the it is useful to recall the despair that gripped much of the nation a century ago. As H.W. Brands points out in some respects the 1990s resemble the 1890s.

> Myriad mergers, acquisitions, downsizings, outsourcings and off shorings creating anxieties, America's exposure to the whims of financial fortune recalled the conditions of a century before. In each decade, the stomach-churning changes, economic and otherwise, evoked a variety of attempts at accommodation, amelioration and denial...fear and resentment pervaded the politics of both decades. Each time, the sense of sliding over a cliff into an alien future prompted unprecedented attempts at heel-digging, and it produced a conspicuously conspiratorial brand of political rhetoric. [4]

> A hundred years later, as Americans approach the end of another century, a single verdict on the 1890s debate between the declinists and triumphalists remains as problematic as ever. As to economic change: The costs of change are usually more apparent than their benefits.... Perhaps more than anything, the experience of the 1890s reveals the resilience of America. The apocalypse seemed nigh during that reckless decade to the many who couldn't conceive that the country could long survive the political, economic, racial and cultural forces that, were tearing it apart. Survive it did, however, and indeed emerged more buoyant than ever.[5]

History has been kind to America. We can hope that it will continue to do so. Description is not advocacy--or approval. From time to time, dispassionate analysts describing trends and events have been accused of approving the deplorable and even of aiding and abetting abominations. This is most particularly true of analyses of the conditions and prospects for central cities and of the future of slum populations. Portrayals of reality are greeted with cries of indignation. This text is open to this kind of misinterpretation. Description is confused with prescription.

It must be made clear that it would be far better if urban America in the late 1990s were more like the cities in the 1950s: confidence in the future, much faith in governments, more trust in corporate employers, thriving and bustling housing markets, crime rates low, schools good to acceptable, growing job markets and near certainty in a better future within city boundaries. True, there was rampant racism, a vicious red scare and a pervasive fear that confrontation on the China coast, in Korea or elsewhere would erupt into all out nuclear war, but, on the whole there was hope in the relatively civil, livable and hopeful societies. The fact that the cities and communities in all dimensions have decayed since the 1950s is an unfortunate fact. Reform, if it can be effected, will be slow. In time the savagery may pass, predators diminish, schools revivified, housing and employment markets recover. What we do know is that technology has very little to do with altering deep seated cultural practices.

For example, in the 1960s we hoped that
- teaching machines, computer learning and Sesame Street imagery would greatly improve educational prospects for slum children,
- surveillance cameras, metal detectors, scanning devices, computerized data bases and advances in forensic science would reduce the crime rate, and
- new contraceptives would reduce out-of-wedlock births.

What we found is that new technology made very little difference. But we also discovered that although technology has only minor impact on serious social problems, it does help us to walk away from them. In the 19th century wealthy people used ferries and commuter railroads to get away from city environments to homogeneous, green suburbs and later in the 20th century streetcars and automobiles offered the same kind of residential options to the middle class and working class. Some may choose to stay and fight with crime watch organizations, work with community development corporations and churches, engage in efforts to reform city governments, schools, police and public services. Good luck to them. They have chosen their option. The great majority are numbered in the ranks of the dissatisfied. Most are existing or potential relocators fleeing from bad communities that will likely get worse, to the better communities that may even get better.

In the 19th century the major changes in technology that gave people a choice of residence were mostly in the field of transportation. When no one but the well-off could afford to own a horse, we had walking cities where workers had to live within a mile or two of their work place, i.e., 30 minutes to an hour's walk. The horse-drawn omnibus running on streets was followed by the horsecar running on fixed rails. Then came the commuter railroad and the streetcar followed by the automobile. Each invention was responsible for a two mile increase in the distance between home and workplace.[6]

So far as travel is concerned, we can expect increasing resort to teleworking to diminish commuting time, cost and effort, some of it in the home, probably most of it in small suburban centers. A surge in telecenter construction is on the near horizon.

In support of the axiom that to those that have more is given, upscale suburbs (and their exurban clones) have a leg up in the amenities wars. They are out in front in the competition for the home based businesses and local centers that will function as the industrial incubators of the future. To exploit their advantages requires only minor adjustments to zoning regulations in single family zones and restraint in imposing licensing fees.

We can also expect that despite their relatively greater safety than central cities-- crime rates are 80 - 90 per cent lower-- even moderate income suburbs will experience the 'gated community', phenomenon which is spreading throughout the nation: the crime sensitive will pay a premium for the perception of increased safety.

One mundane recommendation: Before widening or otherwise increasing highway capacity on congested routes, appropriate agencies should augment efforts at expanding mass transit ridership, lanes and van pools to explore expanded teleworking to achieve trip reduction. It should not require a crisis like the Northridge earthquake that impelled Los Angeles in this direction. Every area confronting costly highway expenditures might take effective action, particularly if it can find a corporation like Pacific Bell to serve as a partner in replacing a substantial amount of road travel with communications technology.

Clearly there are disputed figures on the current extent of telecommuting-teleworking, and even more uncertainty about the future. At this time we can be sure that there will be a lot more of it and the scale will be sufficient to have significant impact on travel patterns, office space markets and residential location.

The movement to home-based work is counterbalanced by the fact that it seems at least a quarter of these workers get lonely. This is the interpretation we place on the finding that just over a quarter of home-based workers would be interested in working from a neighborhood telework center or satellite office.[7]

The fact is that we will most likely witness a combination of centripetal and centrifugal trends away from many central cities and inner suburbs: more urban dispersal counterbalanced, by small scale clotting in suburbia (especially in outer suburbs) and more growth of recreation areas and college towns.

Telecommunications has moved so rapidly that past predictions, particularly skeptical ones, have proven erroneous. In the late 1980s and early 1990s, for example, it seemed clear that a combination of upper echelon resistance, high cost and still-unperfected state-of-the art held back the potential benefits of telecommunications. Technology failed to reach top executives, and video conferencing was pronounced a flop as a substitute for business travel because the $20,000 systems failed to deliver clear images and audio signals.[8]

In short, despite merciless hyping, desktop conferencing never really made it big by the mid-1990s, but Internet and the World Wide Web are reviving this technology's prospects for corporate America. Desktop video differs from traditional videoconferencing because it doesn't require proprietary hardware or software; images are transmitted over a network to a personal computer. Users have connected their Local Area Networks (LANS) and their Wide Area Networks (WANS) to the World Wide Web.

By the late 1990s we have the beginning of wide scale adoption of video E-mail, video conferences involving two or more conferees and video databases. Images are transmitted through a small (and cheap -- e.g., $300) camera on the top of the PC. Picture quality is approaching an acceptable 30 frames per second. More bridging and routing is still needed but this time predictions of widespread use in two or three years -- by the year 2000--seem fully justified.[9]

For our purposes, telecommuting need not be total. As long as travel is less frequent and decoupled from the twenty-odd hours of peak congestion, we can look forward to less demand for lane space and parking, and quite probably, more and more residential relocation to distant suburbs--and beyond. The amount of peak hour stress on overcrowded highways may substantially diminish. More important, if, as seems absolutely certain, more office commuting will be between suburban and exurban locations, there will be major relief for traffic headaches. Transportation planners would be well advised to rethink proposals for highway construction in this new context. Communications will indeed be a substitute for peak hour travel on heavily used radial highways leading to central business districts and quite probably on circumferential routes as well.

The ranks of telecommuters have been burgeoning thanks to a basic change in the US employment pattern. This shift to temporary and part time work reached record levels in 1994. More than one-fifth of the nation's work force--24.4 million Americans--had only part time or temporary work.

Will the trend, continue at jet-speed? Probably not, but the signs are there for a continuation, albeit at a more measured pace. It is possible that in 50 years we will

> ...look back on our current commuter society as only a blip in history, an artifact of the industrial age sandwiched between the cottage industries of the last century and the information age of the next. As sociological, technological and environmental forces converge, we will no longer need to go to work. Our work will come to us.[10]

At the end of the road we can see more fractures in the society with the relative affluents, 20 percent or so of the population cutting themselves off from people with whom they have little or no personal contact and tend to demonize by virtue of the pathologies served up in the media. Meanwhile, new communications technology simultaneously provides exciting possibilities for learning and entertainment--and the prospect that small, vulnerable segment of the population will be desocialized. Virtual reality will complete a vehicle for the ultimate secession not only from perceived dysfunctions and dangers but from the ordinary ups and downs of real life.

1. Michael P. Conzen, "American Cities in Profound Transition: The New City Geography of the 1980s," in Mohr, op.cit., 282.
2. James Howard Kunstler, *The Geography of Nowhere*, (New York: Simon and Schuster, 1993), 200.
3. Bill Gates cited in *The Economist*, March 29th, April 4th, 1997, 18.
4. H.W. Brands, *The Reckless Decade: America in the 1890's* (New York: St. Martins,1995),3.
5. Ibid, 350-351.
6. Sam Bass Warner, "The Urban Wilderness: *A History of the American City* (New York, Harper and Row, 1972), 85-112.
7. Ibid, 1219.
8. Thomas McCarroll, "What New Age?" *Time,* August 12, 1991, 21.
9. Gene Koprowski, "Desktop Video: Has Its Time Finally Come?", High Tech Careers, *The Washington Post*, October 20, 1996, 47-48.
10. Sherri Merl, "Resisting The Call To Telecommute," *The New York Times,* October. 22, 1995, 14.

INDEX

adative reuse 23-6, 63
air pollution 93-5
Allen, Frederick Lewis 79, 102
Anderson, Sherwood 142, 157
Andreas, Duany 131, 136
Atherton, Lewis 143-4
Austin, TX 39, 45, 67
Baltimore 4-12, 17-27, 31-44, 52, 67-88, 103
Barrett, Jonathan 111
Barzini, Luigi 43, 71
Boston, MA 12, 24-7, 45, 67, 75-99, 110, 114, 132, 142, 144
Boulder, CO 10, 110
Boyz In the Hood 44
Brands, H. W. 210
Brimmer, Andrew 3
Brookline, MA 88, 98-99
Brower, Sidney 53, 75-79, 82-85, 102, 106,
brownfields 11, 12, 67
Buffalo Commons 20
Burgess, Ernest 84
California 104, 117, 122-127, 131, 136, 139, 149
Calthrope, Peter 131
Camden, NJ 17
Canada 34
Carter Administration 5, 40
casinos, resorts 28-29
Celebration 135-138, 155
Charter School 123-4
Chicago Tribune 58
Chicago, IL 14-17, 24-49, 84-99
Cisneros, Henry 14, 40, 46

Cluster Development 104-5, 126, 133, 143, 148
Coates, Joseph 90
cohousing projects 146, 158
college towns 103, 110-112, 151
convention center 26
Conzen, Michael 205
Cornish, Edward 161, 201
crime 17-37, 40-63, 114-24
Dade County 99
Dallas, TX 19
Detroit, MI 15-38, 46-68
development rights 118
Dilulio, John J. 54
Disney 131-135
Disraeli, Benjamim 15
downsizing 21, 33, 63, 64
drug abuse 53
Earthquakes 115
East St, Kiyus 17
Eberstadt, Nicholas 175, 201
economic develpment districts 40, 70
education, basic 2
Eizenstat, Stuart 5
Emerson, Ralph Waldo 83
Evanston, IL 87, 99
Exurbs 151
Fairfax County, VA 92, 98
Falls Church, VA 87
farmland 151
Flint, MI 27
Florida 108, 114-47
Freeh, Louis 55
Fuller, Buckminster 24, 115-16
Fulton, Robert 74
functional metropolitanism 13, 100

Garvin, Alexander	12, 43, 67	Kunstler, James Howard	19, 68, 131, 154, 207, 215
gated communities	4, 22, 50, 57, 63, 84, 85, 133-154	Latinos	34, 40
Gates, Bill	209, 215	Lefrak	35-6, 70
Gautreaux program	14	Levy, Frank	60, 72
Glaab, Charles	75, 102	Lewis, Roger	46, 69, 71
Glasmeir, Amy	63, 72	lone eagles	151-3
Glassman	172, 201	Los Angeles	27-44, 84-95, 139, 140, 148
global cities	41, 64	Loudoun county, MD	92, 137, 155
Graham, & Marvin	64-66, 73, 130, 174	Luddite	163, 170, 201
Great Society	4	Lynd, Robert	143
Greenbelt, MD	85, 102	Madison, WI	30
Hall, Peter	31, 47, 71, 80	Maryland	105
Handey, Blaine	21	Massachusetts	84, 88, 123, 124, 132, 142
Harrison, Bennett	63, 67	McLean, VA	87
Hawaii,	105, 117	Miami, FL	19, 27-38, 45
Henry, John	137, 155	*Middletown*	142
home offices	90-95, 120, 140-154	Minnesota	43, 110-123
homestead program	31	Misgovernment	120-42
Hong Kong	34	model cities	14, 40
Houston, TX	104, 114, 117	Montgomery County	1892
Howard, Ebenezer	108	Montreal	117
Hughes and Lang	87	Moynihan, Daniel	40, 54, 70
Hughes, James	89, 96, 103	Mumford, Lewis	47-9, 71
Huxtable, Ada Louise	136136	Muscamp, Herbert	100
illegitimacy	8, 175-6, 202	Nashville, TN	13
immigrants	2-7, 20, 32-42, 63, 160, 187, 194	National Trust	147
Incubators	109, 141	Nelson & Sanchez	53, 108-9,
infill	104-05, 148	New Orleans	29
infrastructure	166, 167	New York City	15-49, 68-70, 89
interstate highway program	81	Newark, NJ	19, 23, 44, 89, 103
Jackson, Kenneth	74-80, 101-02	Norfolk, VA	23-27
Jacksonville, FL	13	O'Neill, Gerald	170
Japanese National Tourist Org	117	Oak Park	88, 99
Jeter, Jon	10, 67	Oakland	89, 99
job losses	9, 19-25, 63	Office of Tech. Assessment	11, 40-2, 106, 152
Kansas City	81	Ohio	11
Kanter, Rosabeth Moss	45	Olmsted, Frederick	75-77, 132
Kennedy	163-69		
Kentlands, MD	131		

outer suburbs	103-9, 126, 139, 145, 152	smart growth	105, 148
Pacific Bell	89, 96	social darwinism	96
Park, Robert	49, 71	Sowell, Thomas	3
Philadelphia, PA	9, 17-29, 46, 71, 205	Sprague, Frank Julian	77
Phillips, Barbara	84, 96, 102	sprawl	147
Philpott, Tom	55, 72	stadium	26, 28
population trends	63	Stegman, Michael	4
Porter, Michael	62	Sterling, Bruce	166, 199
Portland, OR	1, 112, 20, 39, 43, 57, 120-1, 145	Stough, Roger	105
Prince George County	26, 98, 100	tax refuge	152, 154
Prince William County	119	telecommunications	1-118, 130-199, 205
public schools	58, 113-4, 123-4, 134	telecommuting	15-25, 91, 93, 96
Puerto Ricans	6, 7, 37	teleworking	2, 15-23, 93, 95,
Reading, PA	8, 36, 46	Thurow, Lester	166, 201
Recreation	113-127	Toloedo	27
recreation areas	28, 102, 113-8	Toronto	23-4, 114
Reich, Robert	127	unemployment	2
Renaissance Center, Detroit	20	union membership	163-4, 185, 191
Richmond, VA	77	urban renewal	10, 28
Rifkin	167	utopian communities	145
Rio de Janeiro	50	Vergara	20, 68
Rochester, NY	10	Wahington, D.C.	13-23, 42-8, 61-71, 87-98
Rouse, James	27	Walinksy, Adam	54-5, 71
Rush, David	13-18, 68	Wallis, W. Allen	169
Rybezynski, Witold	112	Warner, Randy	32
Samuelson, Robert	60, 72	Warner, W. Lloyd	86
San Diego, CA	1, 38-9	Weisberg, Jacob	14, 68
San Francisco	4, 27-38, 89, 110, 14, 125, 151	Weiss, Michael	86, 89, 102-3
San Jose, CA	4	Wertenbaker's	144, 158
Savannah, GA	24, 69	Wilson, W. J.	3-7, 39, 62, 67
school achievement test	8	Wilson, William Julius	3-7, 39, 62, 67
Schumpter, Joseph	164	work discipline	2
Seattle	10, 20, 57	Work ethic	6
self-employed	95, 163-169, 174-5	zoning	80-95, 104-10, 146-9
seniors	115-6		
Silicon alley (NY)	22		
Silicon Valley	4, 39, 45		

DATE DE RETOUR	L.-Brault	
2 5 OCT. 2003	0 9 AVR. 2005	
0 3 DEC. 2003		
0 8 AVR. 2004		
0 3 AOUT 2004		
0 7 DEC. 2005		

Bibliofiche 297B